Cultural Vision

Cultural Vision

❖

A Memeplex for the Cultural Evolution

J. Martin Knutsen

iUniverse, Inc.
New York Lincoln Shanghai

Cultural Vision
A Memeplex for the Cultural Evolution

All Rights Reserved © 2003 by J. Martin Knutsen

No part of this book may be reproduced or transmitted in any form or by any means, graphic, electronic, or mechanical, including photocopying, recording, taping, or by any information storage retrieval system, without the written permission of the publisher.

iUniverse, Inc.

For information address:
iUniverse, Inc.
2021 Pine Lake Road, Suite 100
Lincoln, NE 68512
www.iuniverse.com

Cover Design by Rick Newlin

ISBN: 0-595-29146-5

Printed in the United States of America

For my Love,
my Life Partner, and
my wife,

Patty

Thank you
for all your hard work in
helping me to make this book a reality.
Together we are our own tribe.

Contents

Preface ix

Section I The "Roots" of Domination 1
  Humanity's ancient symbiotic relationship with the planetary biosphere
  History of Domination

Section II Effects of Domination 48
  Consequences Domination has imposed on our modern reality

Section III Aspects of Domination 106
  Difference between natural and domesticated social organization
  Features of the vision of Domination
  Institutions and technologies that promote and impose Domination

Section IV A New Global Cultural Vision 134
  Vision in a cultural context
  Memes and memeplexes
  Innate human values as an essential foundation of natural culture
  Living Mythology as an essential function of natural culture
  Foundational cultural dysfunctionality of Domination
  Emerging cultural trends
  A new memeplex to guide humankind in the evolution of new cultures founded upon natural human values

Section V The Cultural Evolution 163
  A meme by meme breakdown of the Memeplex for the Cultural Evolution

Afterword 201
Bibliography 205
References 213
Endnotes 215

Preface

> Culture: "5. a: *the total pattern of human behavior and its products embodied in thought, speech, action, and artifacts dependent upon man's capacity for learning and transmitting knowledge to succeeding generations through the use of tools, language, and systems of abstract thought. b: the body of customary beliefs, social forms, and material traits constituting a distinct complex of tradition . . .* "
> —*Merriam-Webster's Unabridged Dictionary*

Our Planet Earth is Sacred and Irreplaceable

The sweet rich air that flowed through the lungs of our ancestors had to be a marvelous thing. I find the aroma of any dense forest to be almost intoxicating. But what I find most interesting about a forest is how life seems to have a much higher priority than it does in our reality. We might squash a defenseless bug in our buildings, but that act would seem rather senseless in the wilderness.

In our world, we might plow under an acre of forested land for homes and businesses, and view that act as Progress. And it is progress, no doubt, but to what end? Now we see that our corporate culture views the natural world as undeveloped land or resources to be exploited, not as habitat for other forms of life whose right to exist supersedes the temporary monetary value their extinction would provide.

Paleontologists (scientists who specialize in the study of fossils and ancient life forms) tell us that when Europeans arrived in North America, the wildlife that existed was quite different from what we find here today. Whales spouted in profusion in the seas; clouds of pigeons roamed the skies; buffalo herds stretched for what would seem to be forever; centuries old indigenous tribal nations lived in symbiotic balance with and as a part of the natural world.

The forests were so dense and plentiful that a squirrel could conceivably travel from the East Coast to the Mississippi River without ever touching ground. Black Walnut trees were five feet thick. Chestnut and White Pine trees loomed 200 feet in the air. The sweet, thick pungent air of the south was full of magnolias, crab apple, and basswoods. The entire eastern half of America was an enormous unblemished green swath of purity with richly scented fresh air and clean sweet water.

Salmon were so thick in the streams you could literally reach out and pluck one out of the water. The continent was densely populated with animals of all varieties. A rich diversity of plants, fruits, and medicinal herbs grew in great abundance. The cornucopia of life would seem to us as fantastic, as if it were a dream. The richness and variety of vegetation and the animals that inhabited those lush ecosystems would truly seem magical.

Life on our planet, our home amongst the stars that we call Earth, is precious beyond all description. We are composed of its very stuff. We depend on every link in the giant web of life for our survival. It is a one-time endowment that could never be replaced.

Earth is Quickly Being Altered Beyond Repair

Most of us are well aware, though many refuse to admit or acknowledge, that we as a culture are purposefully exterminating this awesome splendor with which humankind has been endowed. Every day we are destroying the rainforests at the rate of an acre a second, day and night, without pause. Every year we destroy enough rainforest to cover the state of Oklahoma.

And not only are we destroying life, we are eliminating the very conditions that must exist in order for the life that we annihilated to renew and replenish. In some cases, this destruction is accidental; in others it is entirely purposeful.

We all seem to be helplessly standing by watching the luxuriant and diverse life that took millions of years to evolve, pass away on our planet Earth. I believe we are now facing certain ecological disaster.

Revelations

In 1994, I read *Diet for a New America* by John Robbins, which was the catalyst in my awakening to the realization that Truth in our society does not necessarily emanate from the corporate sector. It changed me and brought new meaning to my life as I began to wrestle with questions of why our culture, society, and cherished institutions are lying to us.

In the search for the answers to these questions I found more and more evidence of complete degradation of the environment and massive agony throughout the world. But still the reasons for this malfeasance were elusive.

I was shocked at the level of degeneration that was self-evident on so many levels and was amazed to learn how long some of these degenerative phenomena had been occurring. How appalling it was to learn of the scope of corruption, destruction, and pointless agony. It seemed that concepts such as genocide, colonization, pollution, competition, deforestation, child abuse, malnutrition, nationalism, slavery, extinctions, terrorism, poverty, military misconduct, corporate mergers, bankruptcy, mass layoffs, and taxpayer bailouts were not only being manifested in our world, they had now become regular components and features. On and on it went with seemingly no sign of abatement.

Triviality

It also seemed that the typical human experience was filled with mindless, trivial, and repetitive tasks that appeared to be as frustrating as they were boring. I'll never forget going through a freeway tollbooth in Illinois and handing a few coins to an older gray-haired gentleman whose eyes showed a sense of timelessness. He looked like someone who might have been a revered elder in another culture, a valued asset to the society, possessing critical knowledge of life and philosophy. And yet, here he was, trapped in a tiny box, counting coins day in and day out.

The old man in the tollbooth got me to thinking how so many of us, who have so much to offer our culture, are wasting our lives stuck in some meaningless occupation. Why do our lives seem to revolve around jobs and careers? Why do so many of us tolerate spending most of our days in tiny cubicles repeating rote duties day after day, year after year? The work-a-day world just doesn't seem to be a very pleasant way to live the few short years we are given by the Universe, so why do we have to do it? Why do we create such an ugly reality on such a beautiful planet?

Illuminating Interconnections

As I examined the peculiar way we live in our culture, it was easy to get sidetracked on various issues or topics that caught my interest. I had to consciously strive to avoid becoming trapped in narrowing the scope of my studies. I had to continue to widen my focus. It was like using a macro-zoom, or a wide-angle lens. I was attempting to solve the Puzzle of the Big Picture. And as it slowly came into focus, I could see where the missing pieces were. I examined these missing pieces to figure how they linked into the rest of the puzzle.

Amazingly, I learned that all the dilemmas the human race is contending with are interconnected. One malady begets another and another, and sometimes even in self-perpetuating circles. It was maddening.

The unemployment of a single father of four in Chicago may be connected to a slave sweatshop in Indonesia. A child in Nigeria may be starving to death for the same reason a man in Miami has asthma. A dead river filled with pathogenic bacteria in South America may be related to a heart attack in New Jersey. A hurricane's disaster in the Gulf of Mexico may be related to a slaughterhouse in Nebraska.

But most media only report on the individual malaise without linking it to the real root cause. They may examine some superficial links, but rarely do they dig deep enough to uncover the real causative agent.

Acceleration of the Rate of Change

The interconnecting elements that comprise the mounting disaster we are going to dissect in this book are continually feeding the loop, accelerating change. It's not even a debatable point that the Rate of Change in almost every facet of human reality is accelerating. The change in the last 10 years was much greater than the ten before it and so on back in time.

One primary factor for this increasing Rate of Change is the human population factor. The increase of the number of people seems to be fueling the whole system. But that begs the question: Why is the human population soaring in the first place?

Divergent Disciplines

We may understand what impact quantity (overpopulation) might have on quality (of life). What we need to do is accurately understand the empowering factors that are fueling overpopulation. If we can do this, I believe we may better understand how to resolve our conflicts, help our neighbors, allow diversity to flourish, and live in accordance with our own personal values unimpeded by any constraints or imposition of someone else's personal values.

Understanding the myriad of factors fueling overpopulation, the current global culture, and our precarious environmental dilemma requires a little bit of knowledge in a variety of disciplines such as psychology, history, cellular science, nutrition, paleontology, ecology, oceanography, horticulture, economics, sociology, agriculture, nuclear physics, theology, meteorology, anthropology, geography, and political science, to name a few.

Fortunately, I am an officially accredited "expert" of none of these highly skilled fields of study. My official label and function in this world is irrelevant. But if I were, say, an ecologist, whose field of study is a critical component of *Cultural Vision*, my focus on the global cultural dilemma might be debilitating because I would tend to write from that perspective.

In our culture we like labels and mono-functioning slots that define any individual's role, purpose, and worth. We live in a world of the Specialist. A nutritionist might not be able to change a flat tire, and a car mechanic might eat nothing but burgers, fries, and soda. We're taught to focus on a specialty to be an easily moveable peg in the working employee/laborer pegboard. As we fixate on just a small corner of human possibilities, we become dependent upon our experts to tell us what to believe. We have become informationally dependent.

Programmability

I believe that the human creature operates and responds less instinctually than others. That is not to say we have no instincts, because we most certainly do. But our instincts are dwarfed by our enormous capability to adapt to a wide variety of living arrangements that allowed our species to thrive in the tundra, the desert, the mountains, and the jungles.

The human mind is born with the unique capacity to absorb huge amounts of complex cultural data. This data creates our perception of reality. In this sense, the human mind is shaped by this mutual perception, and is "programmed" to accept it.

A child born into a Buddhist family will accept the Buddhist beliefs just as that same child would accept the beliefs of the Mayan culture had it been brought up in that place and time. If you were to approach that same child later in life as an adult and tell them that they lived wrongly, that their belief system was incorrect, they would take great exception with you, no matter which culture they lived in.

I believe this holds true for people in our culture as well. As we live and experience reality as it is colored and molded by our cultural norms, we need to fit in, so to speak. We are taught at an early age what Truth is to our peoples. We are, in a sense, programmed to live in that specific culture.

The innate human tendency to soak up, trust, and internalize cultural norms causes most of us to blindly follow our *perceived* societal authority figures, experts, and institutions, no matter how they achieved their status. Most of us don't have the time, inclination, or the resources to question authority because we are so busy surviving. We can only assume, or at best hope, that they know both what is good for us, and have our optimal interests at heart.

Lack of Accurate Information and Statistics

And regardless of how much time any one of us has had to truly investigate, study, and thoroughly research critical issues that must be addressed, we all come to conclusions and formulate opinions. We may not have all the critical data necessary upon which to base our opinions, but many times people are not aware of this. Most people feel they possess all the information they need to make an educated decision. On top of this, there are entities in our society that actively seek to publicize disinformation to keep the masses confused and unable to fully understand the issues.

Controversial Matters

Complicating this lack of accurate information and statistics is the fact that most of us feel strongly about our beliefs, regardless of how uninformed we may be. Tradition, social convention, and beliefs can be matters that are entirely indestructible. Even a dialogue to simply examine other ways of thinking can upset us. We may cling tenaciously to our beliefs because it defines our identity. We do not care to stand apart and hold unpopular and foreign opinions.

The Big Picture

I understand that we all naturally tend to trust our mother culture—the voice that lives within each one of us, forever speaking to our subconscious minds. In fact, we are genetically programmed to instinctually trust our culture. If we didn't have this trust, or if we lost it, the cohesion of our global societies would crumble and chaos would run rampant. Our species would not have survived and emerged from the tropical forests of Africa and spread throughout the Earth to populate the continents. We wouldn't know what to do or how to act. We need this ability to trust others on a cultural level, because without it we are surely doomed.

We must learn to examine our culture objectively if we are to understand how it functions, or in this particular case, how it is malfunctioning. To do this, however, we all need to get into an imaginary hot air balloon that will help us view our perception of reality from a wider and more distanced perspective that will allow us to see our culture in its entirety.

The only way to see the whole cultural picture is to temporarily remove oneself from it. This means temporarily disengaging all of our psyches from our own personal cultural programming and looking at our perceived reality as a whole from a distance. It is only then that the numerous cultural assumptions we make in our daily lives can be seen for what they truly are.

The Outline

Cultural Vision consists of five sections. The first three are dedicated to creating a complete understanding of our current global culture and humankind's unique predicament.

The last two sections contain the information we need to head our species into a sustainable and humane direction. The concept of "memeplex" is introduced and thoroughly defined in Section IV—but until then, think of it as a cultural blueprint or outline. Section IV develops a mutual understanding of our species, Homo sapiens. Few of us realize how our species evolved and are genetically programmed to function. Section IV also shows how social trends indicate that a new movement towards creating a global culture that is based on true human values is already taking place. Finally Section IV introduces the bigger picture of the changes that must occur to illuminate a new direction for humanity that will herald the next chapter in human history. Section V takes the larger ideals from Section IV and combines them with the tribulations set forth in the first three sections. This creates a detailed roadmap outlining the critical steps humankind must take to create a new sustainable global culture that will reflect our own humanity and cultural needs.

Acknowledgements

Finally I would like to express my deep gratitude to all the brilliant minds whose honorable dedication to their pursuit of truth and knowledge contributed to the information I have assembled for you in this book. I would especially like to acknowledge those individuals whose philosophies, lectures, and writings were my primary inspiration in the conception of *Cultural Vision*: Joseph Campbell, Larry Harvey, Jerry Mander, Jim Mason, Daniel Quinn, and John Robbins.

I

The "Roots" of Domination

> *"In learning to control the growth of plants and animals, human beings demystified nature and set themselves up as her master . . . For primal people, the world consisted of beings, souls, and powers; for the agriculturalist, it consisted of resources and pests."*
>
> — Jim Mason

Three Million Years of Diversity

Humankind has existed and has been evolving for more than three million years. Thousands of cultures throughout the history of our species have come and gone, each one with a unique Vision.

Of the plethora of cultures that have graced the planet, every so often one of them fails. The cause of this failure could be for a variety of reasons: excessive violence or passivity, illness, natural catastrophe, or lack of access to food. Perhaps their culture became dysfunctional for any number of reasons.

While this may have caused these cultural adherents to abandon their culture, it did not impact the survivability of the human race as a whole. Fortunately, back in the tribal days there was great advantage to be found for the human species to have a wide range of diverse cultures scattered throughout the planet: If one or two cultures were to meet their demise, there were still thousands of other cultures out there doing their thing. Those cultures that were successful carried on their traditions and experiments in humanity. The human race continued on.

Co-evolution and Symbiosis

Up to the point in the biological evolution of the planet when the human species appeared, all life had not only evolved in stages or evolutionary steps, but all species—life as a whole—took these steps together. The flowers co-evolved with the insects to make their nectar appealing, nutritious, and available so that the seeds of the plants would then be distributed to bring another generation. Rodents, birds, felines, and the rest of the animal kingdom were all part of the dance of life and balanced their populations by the provision of sustenance to others or by some other means. This is called "symbiosis." They unintentionally cooperated with other species by being who they were and acting in accordance with their own genetic predisposition. In so doing they would keep their populations at sustainable levels and in balance with the rest of the natural world.

The human species was no exception. The plants they ate provided their nourishment. The forests provided their shelter. Each evolutionary step humans made were the result of or altered the steps taken by other species as well. If nature changed, humanity, being a symbiotic component of the natural world, would change too. The human species **was** nature as much as trees, bears, or bacteria. I call this basic foundational fact the **Principle of the Human Symbiotic Relationship with the Planetary Biosphere.**

It is important to the central premise of this book that the reader fully comprehends the fact that humanity is as beholden to the natural physical laws as every other creature. Our physical bodies are physical biological entities, regardless of what set of beliefs or religion one might hold.

Spirituality and Animism

As the human species gradually evolved, it developed a higher level of mental comprehension that gave a new aspect of hominid existence: total awe of the universe and life processes. This new ability to contemplate the greater powers-that-be had an impact on human culture. There was a communion with life and a sense of overwhelming interconnectedness with the land, the plants, and the animals. Spirituality was being a part of life.

Animals were equals, perhaps possessing even greater powers than humans did. Humans imagined that all creatures, even plant life and such inanimate objects as rocks, possessed a life force and perhaps indwelling souls. The life and death of all sentient beings were imagined as being part of the eternal passing back and forth of souls. While they could have no word or label for this worldview, we call it "animism."[1] While we may think of it as a religion, it is really a set of values that merged them with the abundant life in the world around them. All creatures had their place and deserved the same respect as human beings.

There was no distinction between the status of animals as being a lower form of life because they weren't viewed as being non-human. They were seen as being other types of "people" in the sense that we are all equals in the great scheme of things. Individuals from any species might engage in conflict with another for any number of reasons, but there was never any one particular species that would ever take a wholesale stance of superiority over any given number of other species. Life was, as it had always been, in total balance.

The Rule of "Speciel" Equality

Being was perfection. It was enough for any sentient being to just be. No creature, human or otherwise, ever had to apologize for being anything other than who they were. I call this the **Rule of Speciel** (as in "species") **Equality**, and it covered every creature on Earth. The hawk might swoop down and eat the rodent, but that individual rodent lived as it should until that point. The rodent didn't live for the hawk. The gazelle lived as a gazelle should. Maybe the lion would eventually eat it, but that had no bearing on the gazelle's life itself. The human being lived with the rest of creation in this same way. Tribal human beings believed that animals had just as much of a right to be as they themselves.

The Garden of Eden

The roots of our present culture are the result of many singular and seemingly unrelated events in human history, reaching back to, and far beyond from, what modern historians call pre-history. Of course, history certainly did happen then! We've just learned to call it pre-history because our species did not write or record the events of their peoples back then so we can only make educated guesses.

Anthropology, though, has proven that humans have been around for many hundreds of thousands of years. Migrating throughout the world, we populated the continents and created thousands of successful cultures that stood the test of time. Each culture had its own roots reaching back to what seemed to them to be the dawn of time itself.

The oceans, lakes, rivers, streams, forests, plains, mountains, and valleys were rich with diverse forms of life. Coexisting with millions upon millions of species were thousands of human cultures, each with its own customs and traditions derived from thousands of years of evolution. The air and water were pure, sweet, and pristine and the food was wild, powerful, and plentiful. There was no famine. There were no great diseases. Life was good.

People in these tribes lived mostly leisurely lives. They did the work that needed to be done, and the rest of their time was spent doing whatever they wanted to do. Everything was communally shared with the tribe.[2]

The human species existed within tribal cultural societies that inherently lived within the web of life. They only took that which they needed for the present moment from the abundance that surrounded them. They didn't store food because they believed that nature was doing that for them. These cultures viewed nature as the abundant storehouse of all the food they would need.

Their populations were static, neither growing nor dwindling. "Stone-age" people rarely let their populations rise above one or two persons per square mile. In this way they enjoyed a comfortable lifestyle amongst a natural abundance of food and resources.[3]

Grow Your Own

Around 10,000 BCE (Before Common Era), plant life was abundant through an area stretching from North Africa through Iran. It truly was a Garden of Eden of sorts. But when the glaciers receded and the climate changed, this area started to become quite arid. Many tribes followed the retreat into Europe, but those who remained needed to find new ways of feeding their people.

Human linguistic sophistication had made great advances by this time. With this ability to communicate in greater complexity, humans had the tools necessary to invent methods and technologies that could be shared with others. They could recreate natural conditions for their food and concentrate these crops in small areas that could feed everyone.

These people already knew plants grew from seeds. People had spread the seeds of their favorite plants before. Over time, some of these Middle Eastern tribes had come to depend more and more on crops, and less and less on foraging for wild plant foods. In time, they began to take greater and greater control of the land, leading to agriculture as we know it. This was the birth of our culture and has been popularized as the "Agricultural Revolution." This was not a real revolution or an immediate transition. It took place gradually, over many centuries.

Totalitarian Agriculturalism

The advent of towns and agriculture created an abundance of food, more than they could possibly eat. Soon they became several thousand people. They claimed the right to total control over the land and the life of all species that lived on these lands. Daniel Quinn calls this style of agriculture "totalitarian agriculture." I call these people the "Dominators" as they live outside of the natural world and dominate the entirety of land and life.

I call those cultures that continued to live within the web of life the "Animists," as that was their worldview.[4] Animists believe that humanity belongs to the world just as do all the other creatures. But this new culture of Dominators that was cropping up had a different idea. They were beginning to think that the world and all of its creatures belongs to humans—that instead of living on a day-to-day basis, which worked quite well for humankind for three million years, that it would now be advantageous to claim parts of the planet as their own. And in that space, all others would be denied. War would be waged on all trespassers and competitors. This strange new outlook must have been a frightening and maddening peculiarity to all those who encountered it.

The Village

Those areas where plant life was more abundant naturally attracted and could support greater human population densities. And where this abundance was found also brought dependence. Gradually the numbers rose as more people came to rely on community farming.

Because they shared a common heritage of the Domination culture, the neighboring tribes were also experiencing an increase in their populations. It wasn't long before some tribes were finding themselves locked into a particular area by the burgeoning human numbers of all tribes in the surrounding land. This led some of them to intensify their farming techniques to raise more crops to feed more people.

Intensive agriculture did not lend itself to a very mobile lifestyle. Tribes needed to park themselves in order to make this new concept work. Permanent dwellings were constructed in very close proximity. In this novel setting, new human relationships were forged along with the development of hierarchy. Sub-groups and a division of labor evolved to perform the various tasks necessary in caring for such a concentration of people.

Totalitarian agriculture became a very successful mode of feeding populations. It was so successful that those tribes that instituted these techniques grew in size. Instead of being semi-nomadic tribal bands of twenty to thirty individuals, they became villages of hundreds or even thousands of citizens.

The First Agri-Cultural Population Boom

The Domination culture eventually had saturated the Near and Middle East. Domestication of plant and animal life spread throughout these lands and into southeastern Europe as early as 8000 BCE.[5] In this period, the human population soared from around ten million to over twenty million in just three thousand years—from 8000 BCE to 5000 BCE. This is a phenomenal increase considering that it took our species over forty thousand years to grow to just five to ten million members. By 5000 BCE, at the close of this time period, cities of up to ten thousand inhabitants could be found in Egypt and other areas of the Middle East.

The rapid population growth was spurred by the enormous food surplus the Dominators created with totalitarian agriculture. But as their numbers grew, forcing permanent encampments or villages, another phenomenon occurred: With each increase in their populations, they would eventually outgrow their capacity to feed themselves. They would always need more food. To get more food they needed more land. To get more land to feed concentrated populations they needed to encroach upon someone else's territory. And so totalitarian agriculture continued to expand in this way throughout this period: Food shortages wrought the expansion into more land, which would produce extra food that encouraged an increase in population that would eventually cause more food shortages that would start the cycle all over again.[6]

As the Domination culture expanded, it invaded and destroyed all other lifestyles in its path, replacing these independent tribal societies and indigenous cultures with new societies based upon Dominator values. These new societies of Dominators operated upon the belief that all life living on the land they set aside for growing crops or grazing animals is the exclusive property of the human race. All competing species were therefore subject to eradication. Furthermore, war could be righteously fought to claim more land, women, slaves, and animals.

Animals are Domesticated for Human Exploitation

Throughout this 3,000-year period in the Middle East, large animals were increasingly corralled and herded for human use. Goats and sheep were first, followed later by cattle. Hunters became herders. Farmers started to use oxen to pull plows and move the harvested crops, increasing their yields. These early farmer/herders soon learned that favorable traits in the animals they kept could be bred, introducing the concept of animal husbandry. Animal domestication led to the concept of property. They began to imagine that they owned their stock. They believed that all animals were a gift from God. This belief exemplifies their value of human domination.

The Demystification of Female Powers and the Desanctification of Nature

Throughout most of humankind's tribal history women were the primary source of social leadership, if at least considered equals. After all, it was women whom men had always viewed as possessing the powers of life and thus were seen as being closer to nature.[7] But when the male role in procreation was better understood, the female powers were demystified. Furthermore, being close to nature was beginning to be seen as crude and animalistic. Women were dehumanized because of their animalistic functions.

As the human cultural base shifted from living within nature to viewing it as something to be controlled and loathed, the human psyche shifted its point of reference to looking in from the outside. Nature was becoming the great "other." It was these new values that began to erode the spiritual connection humans had always felt with nature and life.

A Shift in Values, a Shift in Deities

Human culture began to develop a superiority complex that was derived from the hunter/herder male point of view. Humans had previously always "worshipped" the natural world in which they lived. Their gods were part of that tangible world to which they felt connected. Bears, mountains, or the sun might be some of the possibilities. With the new paradigm shift that placed the human race at the top rung of evolution, various cultures began to humanize the deities that they worshipped. For many, nature was becoming the force to be tamed and conquered.

The new paradigm was based on the model developed centuries earlier by the hunters.

This was a critical change in the way people interacted with other people, with other animals, and with the environment in which they lived. When there were more people crowded together, human focus turned towards creating personal identity within the society and less with how any individual merged with the natural world. After all, what is present in our life is where we focus our attention. If your reality consists of twenty-five other people living within five hundred square miles of forested land teeming with a rich biodiversity of life, then we think about and interact with that. However, if you live in tiny shacks in a village of several thousand people crammed together, commanding your actions and demanding your attention, then that will be what shapes your consciousness. Here was another way that humans were becoming less conscious of the natural world, because they were becoming more conscious of human relationships. The human connection to the planet was beginning to wane.

Kings and Gods

Humans were crowded upon themselves like never before. As hunting cults that were controlled by males gradually transformed into herding, the ruthless ways hunters must practice to be successful followed and penetrated the societies into which they evolved. Prone to violence because of their heritage, they infected other cultures they conquered with their hostile attitude toward women and nature. This callousness towards life produced warriors who had no qualms about killing other creatures, including humans. These cults had little regard for the needs and concerns of other tribes and societies.

The male dominated hunter/herder cultures whose leaders were kings and would command all respect had great influence in the cultural evolution of the Middle East. As the kings, priests, and warriors wielded greater control over larger numbers of people, they were elevated to higher levels of authority. Kings would rule, sanctioned by the priests, and empowered by the warriors. Belief systems and cultures that revolved around new concepts such as tribal nations, property rights, divine powers, and military might were blossoming and interacting with each other. A king would command the warriors. The warriors would enforce the commands of the priests and the kings. The priest might sanctify a warrior's actions. A king who dies might become a god who is living. Religious, military, and governmental authorities were all aspects of society, each reinforcing and validating each other.[8]

These select men who gained their positions by their ruthless reputations were all-powerful, and so were the deities they worshipped. It was their power and disregard for life that allowed them to dominate.

War Gods

Up until this point in human history, most tribal gods were secondary while the universal deities that would support life, the cosmos, and the world in which they lived were primary. But as the expansion contaminated more cultures, we find tough male war gods reflecting and epitomizing the culture of the desert tribes that were nomadic and had to fight for survival. Each tribe imagined that their god was a sole patron of their tribe. It wasn't a nature-based god that all peoples could access; it was their own personal tribal god. Gods were increasingly becoming more violent in nature.

We know that, previously, the world was certainly big enough for thousands of cultures and excessive warring between tribes was neither normal nor necessary. This is not to say that tribes never fought and were without gods to support this unpleasant and occasional ritual in the course of intertribal relations. But to battle as a way of life was not an innate aspect of the majority of Homo sapiens cultural history; it was the obscene aberration.

But for the desert nomads, war was increasingly becoming a way of life. And so for them, all other gods were increasingly becoming relegated to lesser roles while their tribal gods were becoming more violent. The gods of competing tribes were likewise considered to be evil—and why not? After all, these other gods were champions for their enemies.

Let the Expansion Begin

The human population took only two thousand years, from 5000 BCE to 3000 BCE, to double to about fifty million souls, although other estimates put that figure over eighty million. Although they occupied a lesser land area, the Dominators represented about 80 percent of the global human population.[9] Of course, much of this early growth of our culture took place in the Middle East. There were approximately one hundred thousand people in this part of the world in 8000 BCE, but by 4000 BCE this number had risen to over three million—a forty-fold increase.[10]

(This is also the period that our culture had assumed to be "the beginning of time" up until only recently because of the biblical teachings of creation and the advent of writing, which we will examine later.)

It was during this period that irrigation was invented, demanding greater organization and a centralized power structure. These new hydro technologies resulted in further increases in population. Empires were starting to form as centers of powers emerged in critical junctions.

Due to the burgeoning population, it was imperative that these societies organize and develop tough methods to control, obtain, and protect their food supply. Warfare became necessary in order to protect their territory and the farmlands they claimed for themselves. The first states were formed to manage the armed defense as well as to conquer the neighboring villages and fields. Military technology made great advances precipitating innovations in better and more efficient ways to kill other people. As we know, the race continues today. The first great military advancement was probably the domestication of the horse by 3000 BCE.

The next doubling of our population took only sixteen hundred years. At the end of this next period, which extends from 3000 BCE to 1400 BCE, we find over a hundred million humans, with 90 percent of them belonging to the Dominators.[11]

Diseases were cropping up in areas with higher concentrations of people. This is because high population density was a new concept, and they did not understand the relationship between hygiene and bacterial and viral infections. Excess human excrement and garbage was a new problem.

Irrigation, while providing bountiful harvests for many years in various locations, would often ruin the soils they nourished by leeching out the minerals. Time and time again populations would rise and fall, depending on the success and failure of the agricultural complex that would support each society.

Agriculture Causes Famine

The Domination culture expanded along with the related societies that were based upon Dominator values that were quickly spreading around the globe. The more densely populated these societies became, the more they became increasingly dependent on agriculture. This factor made them proportionately vulnerable to blight, flood, or drought. The problem was that crops that appeared in the same fields year after year attracted pests that thrived on those plants. To feed their people they tried to grow food on land that was prone to occasional flooding, while other farming areas were occasionally hit with droughts. Perhaps smaller populations would have been able to survive any of these agricultural maladies because their size allowed them to move to a different area, or they could find a new food source. But they were stuck where they had dug in.

Another recurrent problem these societies faced was the failure of intensive farming practices. Rather than limit their growth, they would always attempt to intensify their crop yields, which will always lead to depletion of natural resources.[12] The inevitable chain of events regularly brought famine, starvation, cultural collapse, and migration.

Traders and Invaders

Totalitarian agriculture had moved northward and eastward into Russia, India, and China, as well as northward and westward into Asia Minor and Europe.[13] There were mass migrations and the establishment of many rulers, nations, and cities. Various societies developed trade, further encouraging migration and cultural expansion. Buildings and temples sprang up. Nomads from the Eurasian steppes invaded Europe and introduced the use of horses and cattle.[14]

War chariots were built and armies started organizing in order to protect kingdoms and expand their influence. In the Middle East, many groups of disenfranchised peoples were wandering aimlessly. Some were captured and enslaved. Some were raiding other villages. These nomads were viewed with great suspicion and despised as thieves and vagrants. Collectively these groups became known as the Khabiru or Hebrews.[15]

Liaison to the Gods

As overcrowded populations became cramped, many groups began to migrate. Violence and confusion were the norm in those areas that had sudden influxes of strange peoples. Too many individuals were following different drummers and this never before seen collage of strange groups had to organize in some fashion, or perish. To keep the cohesion of the larger communities, priests assumed the responsibility of communicating with the gods. This helped to keep the masses under the control of the elite authorities—rulers, warriors, and priests.

When kings and other venerated leaders would die, social disruption would sometimes be the result. Various cultures would replace a dead king with idols or statues that were revered and worshipped as gods. Priests would hallucinate voices that they imagined were emanating from these gods and would repeat these commands that they were "hearing" as the wishes and desires of the god(s). Religions and Nations were beginning to germinate in the fertile soil of human imagination. The seeds of what we call "civilization" were planted.

Writing 101

Writing was developed to track the flow of products and items in storehouses. As it grew in complexity, they began to record their legends, myths, and history. As writing grew in importance, so did the scribes whose job it was to record the edicts of the rulers and the words of the gods as dictated by the priests. This brought unity and a new inflexibility in culture. Once a law or a rule was written, there was no way to confront it if it seemed unjust, except to violate it. Rulers, priests, and merchants began to utilize writing as a method to exploit and control the masses.

Vanishing Voices of the Gods and the Emergence of Prophets, Omens, Demons, and Angels

Julian Jaynes, in his highly acclaimed work, *The Origin of Consciousness in the Breakdown of the Bicameral Mind*, describes the transformation of the human mind over the past ten thousand years. The two halves of the human brain were fully linked in the human mind of the ancients. This is known as the "bicameral" mind. The left hemisphere of the bicameral mind receives information coming from the right brain that stores and freely shares past experiences and cultural information. The left brain examines the data and acts upon it. This information is not consciously created; it is only what the right hemisphere may perceive as being pertinent to the current situation. This information comes from the vast warehouse of past experiences, which is stored as sounds and images.

We've all heard about right- and left-brained thinking. Ancient humans could do it all. All the information people needed to survive and live full lives were stored in their right hemispheres. When people needed to know what to do, they simply had to listen with their left to their own right brain, which would tell them what needed to be done. When the ancient human was faced with all the things of daily life that required a decision, the answer would come from the right to the left hemisphere as a hallucination, as if the sounds and images were actually being experienced rather than simply remembered. They would hear and follow commands that emanated from their own minds. This full integration of the right and left hemispheres allowed humans to hallucinate, or dream, while they were fully awake. Ancients may have imagined that they were hearing gods, as to them these voices from the past were coming to them from the ether.

But when the population grew, as it was at this time, many confusing events were contradicting the information and direction their right hemispheres were telling them. To survive in these constantly changing times, people needed to disconnect from this ancient pattern of thinking and begin to think more logically, more left-brained, more linear. Increasingly, children of the Domination culture were being taught to think in this new fashion. (Today's textbook style of learning is the epitome of left-brained thinking.)

As a result, people were not "hearing" or hallucinating direction anymore, and were anxious to find help in knowing what to do and how to act in these novel times. Of course, they did not understand that it was their right hemispheres that had been doing all the talking for past generations. Some of them were thinking that maybe the gods had left them because the voices were not as readily available as they once were. In attempting to learn what was expected of them, some subcultures believed that the gods could somehow be divined from the world around them. Bad weather or the appearance of a particular animal could be the gods attempting to communicate, in lieu of actually speaking to them. Astrology as a method of divination marked the advent of astronomy. Dreams were believed to be omens. And if omens weren't readily available, sortilege was used, such as asking a question and then casting lots to determine the answer.[16] (This was the invention of games of chance.)

And so began the search for and the study of the Divine—that eternal wisdom and voice bestowed upon the members of any given tribal unit that had existed for perhaps thousands of year in balance and stability. It was those mystical all-knowing spirits that always provided the answers for all dilemmas and predicaments that life presented. As this new more linear global consciousness was developing and spreading, the inner tribal and personal voices of the gods that were previously depended upon for guidance were beginning to vanish. Humans could no longer wait for that inner voice to come out of the ether to instruct them because their tribal culture had no previous experience with these new situations that needed new strategies. Humans looked to special people who might divine the gods and give the masses the instructions they needed to live.

Widespread Chaos and Calamity

The next doubling of the human population took only fourteen hundred years. Between 1400 BCE and 1 BCE, humanity had bloomed to two hundred million souls, and 95 percent or more of them belonged to the Dominators, in Asia as well as Europe and the Near East.[17]

The practice of domesticating and herding large animals was unique to the Middle East. This was a new concept for humanity's relationship with life on Earth. As the human species was becoming artificially sequestered in tight corridors as they were here, males were becoming herders rather than hunters. They lived with their herds and traveled with them. Some of them developed trading relationships with the farmers. Some of them hated the farmers and settlements that were spreading out, pushing them and their herds off traditional lands. The conflicts intensified around the issues of who could righteously lay claim to territory.

Those villages that ruthlessly stripped the soils of the lands around them would eventually experience mass famine. They could not see the gradual changes that herding large animals such as goats, sheep, and cattle were perpetuating on the surrounding land. The changes that would always bring about a catastrophic ending would be imperceptible within one generation. They could not understand that when large animals are concentrated, these animals would eat the young seedlings and trample on exposed soils. Fields where these creatures grazed eventually turned to dust. The human populations these agricultural lands once supported would have to become nomadic, as humans once were. But now the Middle East was overpopulated. There were too many people in this small area, and returning to a forager lifestyle was no longer possible.

The Aryans and Celts (Indo-Europeans) swooped down from north of the Black Sea to invade Europe. These warrior people were patriarchal and were hunter/herders. As they conquered the agricultural societies of Europe, their cultures would first clash and then would merge. This was also happening in the Near East with the Semites invading from the deserts.

Refugees from various areas were vying for homelands and finding no room left to call their own. Huge tribes were intermingling with others and finding that other groups behaved quite differently from themselves.

Agri-cultural Myths

There was a great deal of movement of people in the Middle East during this period. This was due to trade in various commodities, as well as migrations of entire tribes due to famine caused by poor farming practices or the destruction of arable land by domestic animals. As communication between various societies increased, ideas and folklore were shared. As people moved about, they carried with them their fables and legends. This infusion of one set of beliefs into another eventually gave rise to a universal acceptance, over a broad area, of basic foundational beliefs.

The bond that was most common to all these people was totalitarian agri-culture. Therefore, any myth or set of beliefs that supported this common practice would be substantiated and held to be a mutually accepted truth over a broad range of societies.

Previously, when human cultures would evolve anchored in a specific location, certain places or landmarks would become holy places. A meadow, a waterfall, or a hill might have special deep historical and spiritual meaning to a specific tribal culture that evolved in that place. But with so many cultures becoming nomadic, they needed a portable spirituality that didn't anchor them to a specific homeland. It wasn't a place that would be sacred so much as it might be something else not quite so concrete.

The earth itself, along with the animals that ran free, was also becoming not so much a place to be regarded as hallowed ground, but rather another aspect of existence that needed to be mastered and put under human dominion.

What sort of myths would farmers and herders cling to? Perhaps those that gave them the divine right to do what they were doing. They would need to be divinely sanctioned to control and manage the land and the animals. They would need a god that allowed them to appropriate the lives of animals for their own personal gain. They needed a mythology that provided an origin of this freshly evolving culture.

Nothing was unusual about this except for three things we have already covered:

1. Many societies were crowded up against each other.

2. This brewing caldron of humanity practiced two previously unheard of methods for feeding themselves: totalitarian agriculture, and large animal herding and domestication (animal husbandry).

3. Ideas and laws were becoming written for the first time, as opposed to the use of oral traditions for the transfer of knowledge to the next generation.

The third point is critical because with the advent of writing, successive generations didn't have much room to fashion the traditions and customs to adapt to new conditions as had been the case for thousands if not millions of years. Mythologies, once written, could not live and evolve because they remained anchored to another time, or even another place and another society. With writing came an inflexibility that haunts us today. How could a human culture advance with an anchor tying it to one specific point in its evolution?

The Agri-cultural Creation Myth

At the foundation of Agri-culture is the basic tenet of Man as the inheritor and ruler of all the Earth and creation by the Creator. These were the prevailing beliefs by 800 BCE, when the book of Genesis was written.[18] Several tribes had come to believe that the Creator gave them all the Earth to care for.

The deities these groups worshipped were patterned after that which was most familiar to them—herding. Their gods were the good shepherds of their people. They likened themselves to the animals they herded. They then designed their gods to give them the sacred right to be shepherds to the animals. As the gods cared for humans, so should the humans be good shepherds to the animals. That care, known as stewardship, also extended beyond their herds to the earth itself.

Rules, principles, and holy commands were passed in written form from several generations, and eventually coalesced into what we think of as religion. There then sprang many variations of the primary belief in a fatherly all-powerful deity that protected His flock and sanctioned their actions. Present day Islam, Judaism, and Christianity all have their roots in this ideology.

Some of the historical myths that we are about to discuss are alleged to occur before the period we are presently covering (1400 BCE—1 BCE). But we are discussing them here rather than earlier in this book because these myths of cultural history were invented in this period, and do not reflect actual physical events.

While this book, *Cultural Vision*, is not intended to focus on religion, it is important to remember that it is the religious beliefs created during these times that have shaped the global culture we live in today. These beliefs, though many people today may question their literal meanings, have nonetheless shaped and created the current global culture. This is why we will persist over the next few pages in exposing and examining the roots of these religious systems so that we will have a better understanding of the culture they have created, which we are living in today.

I'd like to add that my objective is to be just that—objective. It is absolutely not my intention to disparage the good people who believe in or follow the teachings of any particular religious order or creed. When I began my research, I was actually somewhat surprised by the extent to which certain religious systems exerted influence on the evolution of our culture.

The Story of Creation and the Fall from Grace

The origin of this most famous story that is the foundational creation myth for the Domination culture is hotly debated. Let's review the story itself (which I've paraphrased to make it easy to read) and then examine three different ways of interpreting its meaning.

As the story goes . . .

> *God created the heavens and the earth and filled it with animals and plants. Finally he created Adam and then Eve. They lived in a beautiful place, filled with abundant life, created for them by God where they could live forever in peace. The serpent lured Eve to the tree of knowledge, which was strictly prohibited by God. The fruit that was so tempting, as the serpent enticed Eve, would make them as God, for their eyes would be opened. Having eaten from the tree, Eve seduced Adam to eat it too, and having done so, knew they were naked. But they did not know that they would be banished from the Garden of Eden. God punished Eve by making Adam her master and forcing her to bear the pain of childbirth. He punished Adam by cursing the soil and telling him that all his life he will struggle and sweat to master it to survive. God sent him out to farm the land.*
>
> *They bore two sons, each choosing a different way of life. Cain became a farmer and Abel a shepherd. When it came time for God to accept their offering of the fruits of their labors, God accepted only Abel's. When Cain became angry, God warned him that he could be bright with joy if he wanted to. God said that sin is waiting to attack him and he could conquer it if he would obey.*
>
> *But instead, Cain murdered his brother out on the fields of his farm. As punishment for this horrendous crime, God said that Cain's land would no longer yield crops and that from now on he would be a tramp upon the earth, wandering from place to place.*
>
> *Cain's son Enoch founded a city. Some of their descendants were cattlemen, musicians, and foundry workers and later began calling themselves "the Lord's People." The crime rate rose rapidly and Man became the scourge of the planet. Cain had many children and ancestors that were as rotten as he was and they became criminals and overpopulated the Earth.*

Interpretation No. 1—Literal

The creationist view is that nothing existed before this point in time. Creationists profess that about five thousand years ago God created the Universe and Earth and the heavens and all of the creatures in the space of seven days, and finally created two people in His image which he perched at the pinnacle of it all. Being the closest thing to God, the implication is that the human species was closer to the Creator than the rest of the animal kingdom. The story of Noah (which we will examine next) perpetuated this belief. But having just as much of an impact was the belief that mankind was placed in a position that was closer to God than womankind. This is because most of these cultures were based upon man's work: hunting, herding animals, and warfare.

Interpretation No. 2—The Advent of Agri-culture[19]

In this view the apple represents totalitarian agriculture. The Garden of Eden is the natural world. Adam and Eve represent the ancestors of all of humankind before the Domination culture. Once converted to Domination, banishment from paradise was the price they had to pay as they now were forced to sweat, toil, and labor in the fields of totalitarian agriculture. Cain (Dominators) killed Abel (animists) because Cain had to forever expand to feed his people, even though God warned him that he could be happy if he changed his ways.

Under this interpretation, the story of the Fall originated amongst the animists who were pushed off their traditional cultural homeland. Abel represents these nomadic pastoral indigenous peoples whom were loathed by our aggressive cultural ancestors. The descendants of the agriculturists, represented by Cain, preserved the story of their wild adversaries without fully understanding the true meaning of the story. Eventually they adopted the story as their own creation myth. Later it was recorded and the story was taken literally.

Interpretation No. 3—The Loss of Innocence[20]

This interpretation of the creation myth explains the loss of innocence that was beginning to be experienced by thousands of people through an altered state of consciousness. Adam was deceived and suffered because of it by the "consciously" scheming Eve. Humans were becoming responsible for their own conscious actions because everyone in any individual's realm of influence was no longer a close tribal member that was trustworthy and had the same values and beliefs. They were forced to think in terms of linear time with themselves as objective players within that context. This is the exemplification of the breakdown of the bicameral mind covered earlier.

Animals, as well as tribal humans, lived for the moment and reacted spontaneously to stimuli in their immediate environment. Deception, however, requires the ability to narrate an imagined set of events along an envisioned spatialized linear time scale. Humans that lived in small tribes had no need for this type of rational thinking, and did not develop it. Humans previously, quite simply, existed.

Without the stability of the cultural vision that tribal cultures had collectively provided humanity previously, humans that lived within or were influenced by Dominator culture needed and were enabled to construct and narrate themselves in the world around them. Humans, once they had eaten of the tree of knowledge, had their proverbial eyes opened and "knew that they were naked." They could see (imagine) themselves as others saw them.

The story of the Fall is a fable of new awakening of global human consciousness and self-awareness. It is a fable with a moral that illustrates the newly acquired ability for humans to deceive each other, and why it is destructive. It was the lowly snake and then the woman Eve who practiced intentional deception.

The Hybrid Interpretation

I feel that the truth may lie in all three. An advanced consciousness would be necessary for the snake to purposefully deceive, and for Eve to conspire. Yet the concept of agriculture as a force that was harshly dealing with neighbors is also very apparent.

Finally, all cultures have a fantastic mystical creation myth wherein their particular cultural vision and the values that support it can be deciphered. By looking closely at the meaning behind the myths, the cultural vision can be distilled. Creation myths give any culture a sense of who they are and why they are here.

But this culture was different, as it was destined to wipe out all others. It's our story; it's our fate. We imagine ourselves as being descendants of Adam and Eve. Our culture is still acting on this basis, and as we will soon see, had and continues to have a devastating effect on our own sustainability.

Noah's Ark

Another one of the foundational stories or myths that helps to create the basis of our current global cultural vision is also found in the book of Genesis. God, as we have seen, has given all of creation to Adam's ancestors. One of these ancestors was Noah. The story of Noah is a reaffirmation of man's place in the sacred scheme of life on Earth—at the pinnacle of it all. In this story we find . . .

> God has become quite displeased with that which He has created. The people have turned their backs on Him and developed sinful ways. So, to fix this unfortunate situation, God decides to destroy all life and kill everyone, save Noah and his immediate relations. To Noah, God has left the responsibility to build a vessel capable of carrying a breeding pair of each species of animal. And after God floods the entire planet, Noah saves the animal kingdom from extinction.
>
> God has Noah bring a few extra particular animals specifically for food and sacrifice to Him. He warns Noah to drain the blood from any meat he may eat. Noah along with his three sons and their descendents are to repopulate the Earth when the flood is over and free the breeding pairs of animals.

It's clear to see, no matter how one may interpret the message contained within this myth, that it is the intention of the writer to impart to others that God once again has given Man the sovereignty of the planet and all its creatures. While God Himself may be most high, it is with His divine sanction that Man rules Creation. God also has irrefutably shown that Man is superior to Woman once again—that the only godly reason Woman exists is to enable Man to create his own patriarchal lineage.

With Adam and Noah thus maketh our culture's mythological creation story. But before God seemingly decides to come no more and the foundational mythology is sealed, the authors of the Bible have God intercede in the affairs of his human creations several times over the next few thousand years.

According to the Bible, one of Noah's descendents, many generations later, was named Abram, whom God later renamed Abraham. His story is the beginnings of great new nations and the seeds of a new people. The following account is again, a paraphrasing of biblically recorded events:

> Abraham was a chieftain of one of the wandering desert tribes known as the Hebrews. Around 2000 BCE he led his tribe from Mesopotamia toward the Mediterranean. He traveled to Egypt with his wife Sarah, who also happened to be his half-sister. They pretended not to be married so the Egyptians would treat them better and not kill him to get her. They took his wife into their harem. When God sent a plague, they realized they had been fooled and banished Abraham and Sarah who then moved north near a town named Bethel.
>
> It was there that God promised Abraham that he would become the father of a mighty nation, even though his wife has passed menopause. Since she was not able to bear a child, Sarah suggested that Abraham have a child with their Egyptian maid. Then she got jealous and beat the maid, who subsequently ran away. However, God told the maid to return, which she did, and she gave birth to a son named Ishmael (who later is credited by Muhammad in the seventh century as being the ancestor of all Muslims).

Abraham's nephew, Lot, had two daughters. When Lot's wife was turned into a pillar of salt, his daughters got him drunk and had sex with him. They both got pregnant and their sons were the progenitors of two nations, the Moabites and the Ammonites.

Abraham sat and shared a meal with God as a friend, (something no one else has ever been able to do). *Finally, after enduring a nomadic existence, Sarah does fulfill God's promise and bears a son named Isaac whom God asks Abraham to sacrifice in what turns out to be a test of his loyalty.*

The Emergence of Yahweh (a.k.a. God, El Shaddai, Allah, He-Who-Is, Elohim, and Jehovah)

It may seem a bit confusing that the emergence of the God we know of didn't actually happen until after the mythological events we just covered supposedly occurred. This is because He was written about and imagined at a later point in history, which we are about to cover. What and who the people previous to these times actually worshipped could not be the same deity that emerged as the one God of Judaism. As we shall soon investigate, it wasn't until Moses that the actual deity we call "God" made his first of many earthly appearances, later reinventing Himself according to the times.

Way back before this period, between 2000 BCE and 1200 BCE, three waves of immigrants settled in Canaan (which is now modern Israel). The third wave brought Yahweh, an exclusive new tribal god of war, to these tribes that were fighting for their survival. Moses, supposedly a descendent of Isaac, was a warrior chief in this federation of tribes. Part of the Bible written around 600 BCE is of these migrations. But is this a different god than Abraham knew? By now it seems that God has changed. Moses must hide his face from God who now apparently has decided to appear in the form of a burning bush.

God's personality is always changing and appearing in different forms. While we later learn that "He so loved the world," the murderous god of the Old Testament seemed rather cruel and sadistic. He was certainly partial to just his favorites and extremely brutal to their enemies. He destroyed his rivals, and put men in charge over women. He encouraged prejudice and hatred of the nonbelievers. He also sent various plagues on the people of Egypt such as locusts, frogs, bloody rivers, impenetrable darkness, and the worst one—death to all first-born sons.

What kind of god is this? Cruelty for what would seem to be sport. Yahweh hardens a pharaoh's heart so that he won't let his people go, giving Yahweh another opportunity to kill more people with plagues. Then he drowned an entire army.

Yahweh was apparently not only cruel, but he was selfish too. He commanded that He was the one and only God, and all who did not worship and obey Him would perish. This is a major component of all the religions that emanated from this core value—they were totally intolerant of any other systems of belief on land they considered or coveted as their own. Eventually this intolerance even extended to situations when the basic deity was the same and/or societies on other lands or continents. Domination is not tolerant.

The Biblical Endorsement of Ethnic Cleansing

Historical record of some of the wholesale annihilation of entire groups of people, including the barbaric slaughter of innocent people at the behest of the Lord so the Lord's people can invade and conquer, can be found in the books of Judges and Joshua. Back and forth these groups spawn hatred, distrust, racism, and cold-blooded murder of each other. All of this with a wink, apparently, from the Almighty God of the entire Universe! For this the pious can only say that God works in mysterious ways. Mysterious indeed! This type of warfare has been the hallmark of the global Domination culture and "civilization" ever since. The only way it seemed that potential foes could be dealt with was (and is) to completely wipe all of them off the face of the earth. The Old Testament, arguably the foundational document for our culture's morality, speaks clearly of incidences where bloodshed and wholesale murder are represented as being righteous deeds.

The Diminution of Animals, Women, and Sexual Desire

Increasingly, animals were looked down upon as lower forms of life, beneath the dignity that God had given men. They were no longer considered to have souls, as they had been previously. People tried to disassociate themselves from animals by denying any resemblance to them in their own selves. Bodily functions that were shared by the rest of the animal kingdom became embarrassing and disgusting. Sex and sexual desires were shameful, causing another diminution, that of women. So as nature and animals had to be subdued, so did those aspects of humanity that harked of natural biological processes. They needed to deny the connection in order to feel superior to the rest of creation.

Animal husbandry also affected the agrarian attitude of sex. Sex was for multiplying the herd, and selective breeding was to distill desirable traits. Now human sex was likewise considered to be just for the purpose of increasing a man's wealth by how many children he had to help him with the chores and carry on his lineage. Any sex outside of the specific purpose of marital breeding was sinful. Other normal human sexual behaviors such as homosexuality, bisexuality, masturbation, polygamy, and promiscuity, were becoming punishable offenses by the religious or governmental authorities or institutions.

To the male theologians in almost all positions of authority of the new religions, women were increasingly seen as being sinful adulterous temptresses, here to lure men into loathsome alliances. They were inferior and had to be put into their place by serving, obeying, and surrendering themselves to men in all facets of their lives. Even today, a male hierarchy on a global basis controls religious authority, as well as business and politics. Men are also expected to have control of their wives. The nasty business of war and the technologies that support it were all in the realm of the male. This also contributed to the male domination of hierarchy within Dominator societies.[21]

Nations, Power, and Repression

Centuries old ideas, languages, information, and traditions were being wiped out as the onslaught marched on. Trade between nations increased. The human species was beginning to alter the landscapes as temples, palaces, mosques, cathedrals, and shrines rose throughout all of civilization. Counterfeiting, inflation, and economic inequality were commonplace. Civil revolt, slavery, and military action grew as the global society became more polarized, created by class struggle.

The religions became crutches for the working class and began to promote the concept that suffering is part of the human condition. They taught that the reward for suffering is found in a new life after death. The reward is given to the believers for living in accordance with the laws of God.

This misery was not contained in just the Middle East. In the Far East and India, Buddhism was spreading the concept that all life is suffering, and if one is enlightened one can attain nirvana, the state in which one is impervious to the lower human emotions. The Buddhist's goal is to be unmoved by and completely detached from earthly feelings of greed, hatred, and ignorance to attain the rapture.

Many empires rose and fell in this period ending with the Roman Empire controlling a vast portion of the western world around the Mediterranean. A huge trading network was created to supply the Romans with their wealth.

The Suffering Increases in Size and Intensity

The next doubling of our population would take only a scant twelve hundred years. At the end of this period of war and conquest, from 1 BCE to 1200 CE (Common Era), we will find four hundred million humans. At this point, 98 percent of the human global population belongs to the Dominators.[22]

This incredibly dark period in human history brought war, political corruption, crime, civil unrest, mental anguish, increased human servitude to the ruling class with greater economic disparity, the Roman Empire and its collapse, and the invasions of Europe by military powers from the Mongols, Goths, Normans, and Moslems. Slavery became more prevalent as class distinction became increasingly marked. War within the powers of Europe was a constant reality as the entire continent descended into the Dark Ages.

As the populations in urban centers increased, so did the overcrowding and poor sanitation. This was misery hitting a new low. Famine swept throughout the land along with the plague. Riddled with diseases never before seen, the human condition continued its decline. Mass discontent was spreading throughout the more heavily crowded population centers.

The Christ Cult

Jesus was a man who, at the age of 30, began to preach that the Kingdom of God is at hand, which he taught was something that was available for all people, not just the Hebrews. With his message of universal brotherhood of all humankind in the eyes of God he became very popular, and he initially attracted a large following. The Jewish leaders, not wanting to share their authority or status, denounced him to retain their power. His popularity dwindled and three years later he was detained and handed over to the Romans who crucified him as a revolutionary. His teachings were preserved and he became posthumously famous and the myths began to propagate. Soon Jesus became the Son of God himself and was worshipped as the Christ who died for our sins and our personal salvation, as long as one believes in Him.

But the first followers of Jesus thought of him as only a teacher.[23] They were not aware of any of the incredible events that are elaborated upon in the gospels. Mark wrote the first gospel around 70 CE, followed by Matthew, John, and finally Luke.[24] The mythology behind their writings borrows heavily from Greek, Jewish, and Mithraic myths and ceremonies.

The similarities are many: The bread and wine of mass was a ritual already in place with Dionysus, son of Zeus, whose blood was drunk as transubstantiated wine.²⁵ The Mithraic cult drank the blood of their bulls. December 25th was their holy day. The Christian Church chose the bull, the idol of their competition, to symbolize the devil.²⁶ The concept of the son of a god being heir to his father's kingdom is ever present in Greek mythology, as is the tradition of dying a noble death. Easter was originally a springtime pagan fertility celebration. Eggs which represent fertility and the prolifically fertile bunnies are obvious ties to this ancient tradition, as is even the very name of the celebration: Easter is named for Teutonic goddess Eostre, the Goddess of Spring. Sunday was the day of the week the Romans set aside to honor their Sun god.

There was much debate in the early Christian theological circles as to whether or not Jesus was connected to Judaism at all. They debated whether or not it was Yahweh who was the Father of Jesus, or a higher power. There was a great effort to tie the Old and the New Testaments together so that the prophecies in the Old could be seen as being fulfilled in the New.

As Christianity declined in the place it originated, it flourished as Constantine and then Theodosius adopted it as Rome's official religion toward the end of the fourth century. When the Roman Empire collapsed, Christianity spread throughout Europe burning with evangelical fire, and would be the catalyst behind the soon to come expansion and spreading of the "Good News" to the unsuspecting indigenous societies yet to be encountered.

The Jews were suddenly the minority group throughout Europe as religious persecution grew. They were unfairly taxed and severely discriminated against by the Christian authorities and other institutions that fell under the influence of the Vatican. Then, as now, the religious and state institutions coerced their subjects to believe or be persecuted and perish.

The Great Monotheistic Split

In the year 610 (CE) an Arab merchant in the city of Mecca named Muhammad ibn Abdallah became convinced that he was receiving revelations from the god of Moses. He felt that the city was moving away from God and worshipping materialism instead. While Christians were arguing about the relationship of the Father, Son, and the Holy Ghost, Muhammad said that there is no other god but God, and that's it, period. The simplicity of this concept spread through the Middle East and stretched from Spain to India. His revelations were written by scribes and collected together some years after his death. Such was the birth of Islam.

Salvation

Sages brought news of other worldly pronouncements and laws. Religious leaders became ever prevalent and tried to show their people the way to salvation and a life everlasting. People craved assurance that, upon death, they would live in a heavenly state of peace and tranquility and receive a new life without the pain and suffering that they lived with now. All they had to do was have faith in their deity. Salvation would be their reward for their unquestioning faith.

Personal salvation became all-important. The ordinary people were anxious to escape the life of the downtrodden and dreamed of a Heaven filled with peace and tranquility, where the meek shall triumph over the rich and powerful ruling class. Moslems, Jews, Hindus, Buddhists, and Christians alike all taught the road to Salvation to the desperate masses.

All the new religions taught that whatever the current conditions of society were, that it was human nature for it to be so. After all, the original sin (committed by Eve, thereby implicating all future women as potential sinful vessels) had doomed us to a miserable life. Not only that, but now Christians were teaching that Jesus died for our sins, and that believing and praying to Him for forgiveness wiped your slate clean. How odd it may seem that one can do as one pleases, and believe in a religion that has the capability of automatic absolution by performing a few simple rituals. How convenient to be able to control and blind your own deity like that!

All the new religions taught that nothing mattered but your own selfish salvation in some mystical afterlife. We had developed a miserable existence, and now the only thing people had to hold onto was faith and hope for an afterlife in a beautiful heaven floating amongst angels with God himself. In order to achieve it, all you had to do was believe in it. This concept known as "Faith" persists on today with the incentive of life-everlasting, and without it the threat for eternal damnation.

Personal salvation was all that mattered. Suffering (because we must toil to live) and pain in the earthly world is our destiny, written into our unspoken cultural belief system. Poverty and destitution was the honorable way to live as the meek were promised that the tables would be turned in the next life.

Religions began to play a larger role in the development of our culture. People became fervently obsessed with their style of religion and systems of beliefs. They were willing (as many still are) to back their methods for salvation with their very lives, and then salvation would be their reward.

Myth or Fact

We know these ancient stories well:

- the creation of the first woman from the rib of the first man,
- a talking serpent,
- a god appearing as a burning bush,
- an ark saving a male and female of each and every species of animal from a world-wide flood,
- an immaculate conception,
- a man walking on water, and
- a resurrection of the physical body of a man who is the son of a god.

These are myths, not actual physical events. More than that, these are symbols.

Other systems of worship and honor also have incorporated fantastic stories to illustrate values and practices that could be advantageous for members to learn and understand. But none of them carried with them the zeal to expand or such an air of superiority and arrogance as the god with the capital "G." This is because these myths, which lie at the foundation of today's religious institutions, emanated from the Domination culture. So we find that missionaries carried with them a holy duty to spread the word of their god to all who did not "believe." All must believe.

One of the primary values of the Domination culture is that all of the mythological stories are believed to be accurate and truthful renderings of historic events.[27] Not believing these stories as actual physical events is tantamount to evil itself, and so converting all non-believers was and still is holy work.

Along with that holy work was also the expectation for cultural conversion. New people had to be coerced or strong-armed into the basic ideology of Domination as a cultural practice. The older ways were extinguished, forbidden, and punished. This evangelistic cultural force continues on today, and is still being fought around the world. The consequence is always war and terrorism backed with the compelling logic of patriotic and/or holy duty and obligation. Both holy and patriotic violence are simultaneously honored and condemned in most all the Dominator subcultures, depending on which side you're on.

Famine, Epidemics, and Slavery

The next doubling of our population would only take five hundred years. This period, from 1200 to 1700, ends with approximately eight hundred million humans living on Earth, and 99 percent of them belong to the Domination culture, which now extends around the entire world.[28] Of course, never before has one culture filled the entire planet.

The pure numbers brought on a never before seen flood of diseases and the mass torture and wholesale enslavement and obliteration of entire villages and tribal nations:

- The bubonic plague spread through the continent;

- Genghis Khan invaded Europe;

- Black Death killed millions and devastated Europe;

- Famine struck from England to Japan;

- Syphilis, smallpox, typhus, and diphtheria spread as misery became synonymous with the human condition;

- Africans began to become victimized by the slave trade;

- Inquisitors, sanctioned by the Vatican, murdered and tortured innocent people suspected of heresy (holding a differing view or belief in secularism) throughout Europe and then later in Latin America.

The Dominator invasion into the new lands of the Americas was a catastrophe for every culture they touched. For most peoples, their physical death was swift as up to 95 percent of them died from the new diseases that the European invaders carried with them. This paved the way for the pioneers who thought that it was God who was actually killing these savages for them as they easily advanced further into the heart of the continent. Those hapless victims who were left behind were either cruelly enslaved or brutally slaughtered by the God-fearing Christian aggressors.

For those who could avoid death or slavery, the only other choice was becoming Dominators themselves, either by:

1. fighting these brutish thugs in a futile attempt to wipe them out and eliminate them (something that is a Dominator trait antithetical to animist culture, as well as the Indian tribes of the Americas), or

2. allowing themselves to be assimilated by the invaders, as if their indigenous cultures were somehow inferior and not good enough, even though their ancestors had been living this way for thousands of years.

Beyond that, of course, the only choice left was to abandon their homelands and run, which as we know would ultimately prove to be futile.

The Enlightenment

The great thinkers of this period began to invent the sciences and a new understanding of the world and how life operates based upon quantitative measurements and calculations. In turn, this offered humankind greater control of natural life systems. It also changed the way we looked at where we, as a species, fit in the cosmos. Western thinking began to rethink the role of God, as science gained more respect as an augmented and enhanced study of the Divine.

As new ideas and methods became more accepted and agriculture increasingly delegated to a shrinking percentage of the populace, more people devoted their time to intellectual thought, leading the way to political and social change. These changes altered the way we viewed the human role in the great scheme of things and ultimately their perceived relationship with God.

The irony in the quest for scientific knowledge is that it began as a search for God, to find Him now, as He was no longer coming to us. Humans strove to once again meet God face to face, as did Moses and Abraham just a few millennia before. What we found was not God, but a new left-brained, logical, structured, and secular view of the Universe. New technologies and mastery over natural phenomenon fostered a new attitude that placed humankind where people had previously imagined God to be. While this new attitude still placed humankind outside of nature as we imagined God had created us to be, secularism offered humankind total mastery of life. In short, secularism allows mankind to be as a god.

The Specialists

People began to specialize in various fields of thought. Increased efficiency to maximize technology was critical to the expansion of agriculture, trade, and the military. Human society made heavy investments in the development of technological innovations. Each specialization's advancement would help another do the same, as they all were becoming increasingly interdependent upon each other for continued growth and investment.

New ideas and technological innovation began to spread like wildfire. No longer was cultural advancement in the realm of the elite few, but it was carried out by and dependent upon the laborers and workers, who in turn fueled that growth by becoming consumers in an ever-growing marketplace.

The specialists in all fields collaborated with each other by sharing their discoveries and research. This expanding pool of human knowledge and the pioneering explorers and conquistadors created a new myth in our culture we call Progress. Humankind increasingly believed that they were in charge of their environment and no longer needed to rely on old traditions or revelations from a god to discover Truth. Truth and Destiny were now manifested by Man himself. We were self-reliant and no longer beholden to the natural world.

Specialization Creates Modernism

As the specialists delved more and more into their individual studies, they were also increasingly unable to see the whole picture. A printer knew everything there was to know about printing, was wholly dependant on the local farmers for food. The farmers relied on the tool industry to provide the farm implements to reap their harvests. People focused on their unique contribution to the needs of the community at large and had less time and inclination to learn in any great detail anything beyond the scope of their profession. This made them dependent on the others for guidance in matters of life, relationships, food, spirituality, and technology. This ever increasing institutionalized ignorance of an ever decreasing diversity of range of knowledge is a trend that intensified throughout this period, and continues on today.

These are the roots of Modernism, which was an even further detachment from the true source and from history itself. Moderns accept the urban world as the self-evident way to live—that humans should make a living earning a salary by learning fewer skills but to a higher degree of proficiency. Moderns have blind faith and an almost religious trust in their value system that emphasizes youth, science, institutional control, categorization, and compartmentalization of life.[29] We will be covering Modernism in greater detail later in this book.

The Emergence and Deification of the Nation-State

The idea of an empire-nation was well established in the civilized human psyche by this time. The human need for structure, as we had evolved within the social tribal structure, was now supplanted with allegiance to a greater, less personal, national complex.

Sacred oaths of office now sanctioned the authority of the political leaders. National borders, no matter what peoples were existing in the areas where they were drawn, became new human boundaries that were supported by the authorization of the religions and imposed by the military. Political, religious, and military authorities collaborated in the westernization of indigenous cultures of the world. The diverse community of humanity was quickly becoming homogenized. A singular view of human superiority was spreading. No longer a part of the natural world, humans would occupy land and altered it in any way that benefited those who claimed it as their own.

A sense of the right to exploit the earth emerged. The property rights of the individual and especially of the elite practically became a religion unto itself. Commons areas were appropriated by the wealthy who had the power to control local governments. This new right of individuals to own land transformed into holding the rights of an individual over the rights of the community. If you could legally lay claim to land, it was yours to do with as you wanted, regardless of whether or not it was beneficial for the community at large.

The Triumph of Christianity and the Pioneers of the New World

In many urban areas throughout Europe, anti-Semitic sentiment (hatred of the followers of the Jewish faith) grew as Jews were increasingly viewed as foreigners in their own villages. A prime example of this can be found in the pre-Nazi, anti-Semitic writings of Martin Luther. He wrote in 1543:

> *"What shall we Christians do with this rejected and condemned people, the Jews? . . . I shall give you my sincere advice:*
>
> *"First to set fire to their synagogues or schools and to bury and cover with dirt whatever will not burn, so that no man will ever again see a stone or cinder of them. This is to be done in honor of our Lord and of Christendom, so that God might see that we are Christians.*

> . . . I advise that their houses also be razed and destroyed. For they pursue in them the same aims as in their synagogues. Instead they might be lodged under a roof or in a barn, like the gypsies.
> . . . I advise that their rabbis be forbidden to teach henceforth on pain of loss of life and limb.
> . . . I advise that safe-conduct on the highways be abolished completely for the Jews. For they have no business in the countryside, since they are not lords, officials, tradesmen, or the like. Let them stay at home.
> . . . I advise that usury be prohibited to them, and that all cash and treasure of silver and gold be taken from them and put aside for safekeeping.
> . . . I recommend putting a flail, an ax, a hoe, a spade, a distaff, or a spindle into the hands of young, strong Jews and Jewesses and letting them earn their bread in the sweat of their brow, as was imposed on the children of Adam (Gen 3:19).[30]

But Europe wasn't the only place where Christian superiority conquered and persecuted peoples of differing beliefs. It is important when analyzing the onslaught of the Europeans against the indigenous peoples and environment of the Americas to have an understanding of the type of mindset these invaders had of themselves. These people thought of themselves as pioneers, as God's chosen people. Their mandate was to go forth and fill the earth and multiply. It was their mission to spread the word of their god to all people. Expansion was Destiny, a holy mission from God.

With their cattle, the new pioneers claimed the land as their own, the right they believed had been given to them by God. All these forests were useless unless they were decimated for our culture's fields and cow pastures. "Winning the Americas" was a victory for "God and Righteousness." Righteousness is the morality behind Domination and is exemplified by the Christian slaughter of "pagans, Jews, Muslims, heretical Christians, Indians, Africans, Polynesians, Asians, women, men, children, salmon, forests . . . in the name of a man who said that people should love their neighbors and love their enemies."[31]

Superiority and Isolation

Dominators view indigenous cultures as being evil and satanic. The early pioneers believed that the native inhabitants of the lands they were beginning to occupy were savages, dirty and unkempt, like animals that needed to be tamed. To them these American Indians were subhuman and had either degenerated into a disgusting uncivilized way of life or were of a different species. What many people did not know was that they were coming face to face with how their ancestors had lived in tribal cultures. Our culture has always assumed that human beings have always been "civilized," living in towns or farms while eating crops farmed in fields and livestock raised in pastures. (While we know this is not true, our culture still operates with this belief built into its foundation.)

Dominators imagine that only they, as opposed to other cultures or races, are created in the image of God. Along with this cultural superiority complex they also developed a sort of mass psychosis of separation from the rest of creation. Dominators believe that only human beings, or even specific races or subcultures, have souls and an afterlife. The great religions teach us that spirit transcends over nature and that only our species shares this spiritual capacity with God. So when Dominators think of what spirituality is, they imagine this detachment from nature into some mystical altered universe of another dimension we might call heaven. But as we discussed earlier (and will cover in great detail in Section IV), the true feeling of spirituality is a feeling of a connection with nature, not detachment.

But we don't seek this connection because, according to our culture, we are set above nature. We think that we are so far superior to the rest of the animal kingdom that we have the right to perpetrate any unspeakable horror upon them that we deem appropriate. And the God-given human right to exploitation extends to not just the rest of the animal kingdom, but on all of creation itself. This lack of belonging to the natural world lies at the very root of our culture's health, environmental, social, and institutional decline.

Ethnic Extinctions and the Decline of Indigenous Cultures

The next doubling of our population would take only two hundred years. At the end of this period, from 1700 to 1900, there would be one and a half billion people living on earth.[32] The colonialization of the planet by the European powers was complete as Africa, South America, India, and Indochina were invaded. The Europeans brought new diseases to these areas, infecting the indigenous inhabitants (sometimes intentionally) and thereby drastically reducing their populations. Those who survived the smallpox, typhus, yellow fever, scarlet fever, cholera, influenza, and measles, among other horrible afflictions wreaked upon them, were enslaved, incarcerated, or murdered as the invaders stole their homelands where they had lived for what to them was all of time.

More and more agriculture produced more and more people and this created the modern city. These urban centers of disease and squalor brought new meaning to the word despair. These slums were fertile soil for crime as people who were trapped in these depressing areas were unable to provide food for their families. As crime grew, so did the justice systems that dealt with only the symptoms of crime and not the root causes. The more crowded we became, the more crazy we felt. People were not only getting infected with physical diseases, but people were becoming devastated with mental diseases as well. Many of them were locked up along with criminals, or put in chains. Those people who had lost their grip on "reality" and were fortunate enough to avoid incarceration lived as outcasts in their own communities.

Even though Dominators see great danger in mixing with people who are different, they force their beliefs on other cultures that prefer to be left alone. Dominators do not want their particular race (breed), or subculture to become assimilated or diluted by another. In the Domination culture, each human ethnic race, subculture, and religious sect or cult imagines that only they are the Chosen. And so just as Dominators breed animals, so we also breed ourselves, obsessively desirous to maintain our own pure bloodlines. We can use other peoples for their labor, but we prefer to keep our distance at the same time. And on a larger scale, many populations have been annihilated by mass genocide after they have been dehumanized by a large enough population base with religious, cultural, racial, or national differences.

Institutionalized Diminution

The human species evolved with animals, and so we are programmed to appreciate and relate to the rest of the animal kingdom. But the human genetically innate desire and need to relate to animals was increasingly being unfulfilled as large populations were expanding in urban areas. There's no doubt that humans love and care for their pets. In fact, some pets receive better medical attention than most people do. Humans have an instinctual need to love animals. To satisfy this need in our detached world, we had to genetically alter other species so as to make them compatible with our new lifestyles.

This is how odd pets became very popular—odd in that they were bred with ever-stranger characteristics endowing them with biological deformities to amuse their human keepers. The genetic traits bred into dogs and cats were for the pleasure of human beings and not for any benefit at all of the animals that were created by wholly unnatural means. Not many people are aware that many of these breeds suffer because they are genetically deformed and prone to structural discomfort or disease.

Organizations tapped in on our cultural obsession of feeling superior to other species. Zoos and circuses sprang up for the entertainment of people who became ever more disconnected from the real world, furthering the idea in our culture that animals are strictly for our personal profit and amusement. Seeing animals imprisoned in zoos for us to observe and laughing at the funny antics of circus animals strengthens the notion that we are so superior to these creatures that we can poke fun at them and control their lives and actions. Our status of superiority and mastery over animals is confirmed when we view non-human beings in such humiliating circumstances or under the control, through whatever means, by humans.

The Secular Creation Myth

As the human scientific crusade of research and discovery led to the acceptance of evolution as a plausible and most likely explanation of the creation of life, a new myth of the beginnings of humankind was taking shape.

In the research of the true beginnings of humankind, anthropologists and other scientific researchers attempted to construct their image of what early man was by creating a creature (or evolutionary line) that would be a plausible precursor to the human beings that live within the current world-wide totalitarian agrarian culture. In other words, humans began to see their own species (and still do) only in the context of the culture in which they were living. Evolutionary history was then designed (albeit unintentionally) to terminate in our tainted image of what it is to be a normal human being.

So if we observe crime, social dysfunction, mental disorders, meat consumption, shattered families, sex crimes, drug addictions, and institutionalized competition and violence in the world around us, we think that it must be human to be so. That's why we think of a bloodthirsty, violent, meat-eating ape-man when we imagine the "stone-age" men that must have come before us. They carried clubs and pulled their women around by their hair. They were stupid and lazy. They were brutish and dirty. They only grunted and snorted unintelligible nonsense. We imagine we emanated from creatures that are everything that we fear or find disgusting.

Righteousness is the way to overcome these unsavory qualities of human nature. And yet it is in our own reality, our own Domination culture, that we find evidence of these very traits that cause us to regard our ancient ancestors in such a dim light. In fact, it would seem to us that this type of creature is the only possible forerunner of our kind. The violence that the Domination culture manifests has stained the very image of who we think we are and where we came from. We will come back to this theory in Section IV.

Relationship Intensification

The human species evolved in tribal units that rarely exceeded 30 or 40 individuals, and most human interaction was between those tribal members. Of course, by this period in time (1700—1900), one might argue that people had available to them many times that number to associate with. After all, in the cities there were tens of thousands of other human beings to choose from with which to create a bond.

But there has been a critical difference that is antithetical to human nature, and that is the confining of humans for long periods of time within physical structures. With families, that means just parents and kids. At work, that means a boss with total power and coworkers with different backgrounds. But when that happens, the personal habits of others must be endured and everyone must be careful not to provoke others. Now, with just three or four people to relate to on a regular basis, things that would normally seem trivial can gain great importance; the focus is narrowed and the relationship becomes strained.

Many of the city dwellers were experiencing what I term "Relationship Intensification." Strained social bonds and family and coworker distress were commonplace.

This "intensification" is how families and work relationships started to become extremely dysfunctional. Lacking the loving support system found in any tribal unit within which the human species evolved to live, families were trapped within their own four walls, with only the immediate family members to interact with. There were no subgroups here to deflect the intensity of a one-on-one relationship, no group of women to confide in, no men's groups to spend time discussing the matters of the tribe, and no break from the children. Family relationships took on a much more intensified focus. This claustrophobic living arrangement combined with the poverty-stricken situations that most of them lived in resulted in mass psychosis. Eventually this led to the creation of a new profession—psychiatry.

The Industrial Revolution, Capitalism, and Democracy

The Industrial Revolution was the result of a huge advancement in technology. It seemed that almost anything could be done. Factories were constructed and men, women, and children all around the world were indentured to serve. Barely making enough to survive, they toiled in the mines and sweatshops, slaves to the heartless system created by totalitarian agriculture.

The urban masses were increasingly finding themselves more distanced from their source of food, which was becoming locked up and controlled. In urban areas, the distribution and allotment of food would depend on how well one performed for their masters. No food for you unless you reduce your quality of life to a mindless repetitive chore that would drive anyone crazy. People rebelled, and when they did they were quashed by the military. Tens of millions of people perished in this way during this period.

However, in some cases these struggles brought forth a positive outcome. The right to be able to struggle openly for a share of the power of the state was a hard won liberty in Europe and America. Here the lower classes gained many concessions from the powerful. The rise of democracy is surely the consequence of thousands of years of state-sponsored slavery and oppression.[33]

Up until this period the religions strictly prohibited usury on loans or outstanding debt, but throughout the 1800s the practice of adding interest charges became the norm. Indeed, the Industrial Revolution was funded by usury. This wrought three consequences to the prevailing economic systems of the time, which are now part of how we do business today:[34]

1. **Systemic Competition**—People must return more money than they originally used.

2. **Economic Growth**—Interest requires people to continually earn more than they use.

3. **Concentration of Wealth**—Financial wealth accumulates unto itself as the wealthiest people own and control the vast majority of society's interest-bearing assets.

Capitalism thrives on interest that allows those in power to accumulate wealth through business firms, banks, and control of the land and food production. The system encourages the hoarding of cash, locking it away where no one else may use it. It seems that under capitalism, money itself is the means *and* the objective.[35] In turn, this power, in the form of vast financial wealth, allows those in control of these public institutions to erode democratic rights and freedoms. This is why I believe capitalism and true democracy are antagonistic concepts.

The Emergence of the Corporation

In 1787 there were less than forty corporations, and that number had grown to well over three hundred by 1800.[36] Back then the government and private investors, to serve a specific purpose or perform a special project, could create a corporation. The wealthy saw the advantages of using a separate vehicle for developing income. Their numbers grew.

The East India Company was the British corporation whose oppressive policies, practices, and tactics sparked the revolt in Boston called the Boston Tea Party on November 28th in 1773. Many owners of the corporation were also members of the British government which fashioned tax laws that put other companies out of business in much the same way as Wal-Mart and other large chain or franchise stores do when they enter a community and price the locally owned competition out of business. As we know, the colonies fought back and we call these first corporate rebels "Patriots." (Today, people who complain, struggle, or fight against corporate dominance are labeled "Protesters" or "Terrorists.")

After the Civil War, the Fourteenth Amendment was fashioned to allow all persons of every race full constitutional rights. The reason for it was to free the slaves in the South. But then corporations made many unsuccessful attempts to use this law to apply to themselves as "persons." However, in 1886 a court reporter for the Supreme Court added to the head notes of the landmark case of Santa Clara County versus Southern Pacific Railroad that "corporations are persons" which gives them "equal protection of the laws."

But it was not part of the decision in that court case to give the corporate structure the same rights as a flesh and blood human being; it was just the reporter who decided to add that. History shows the reporter may have had ulterior motives, and there is no doubt now that corporate personhood was never granted by the court. Nonetheless, it has been used in court decisions ever since as justification to equate a live human being that breathes, eats, loves, and sweats with a corporation that employs, can live forever, split into pieces, change nationality, and is merely an agreement that exists on paper and in our imaginations.

This intensified the onslaught on the environment and the human condition because corporations now had the power to control the opinions of the general public with the human right to free speech as guaranteed by the Constitution. As if they were a human being, they could lobby for legislation that would allow them to legally acquire and plunder any land for profit they desire. Only a corporation could have access to the capital to be able to afford huge tracts of land and a mass propaganda campaign to effectively influence public opinion. Individually as well as en masse, corporations have the power to drown out opposing ideas that might be more beneficial to the community at large.

Traditionalism, Patriotism, Global Conflict, and War

Sixty years, from 1900 to 1960, would be all that was necessary to create one and a half billion more humans. It took millions of years to get the human population to one and a half billion, and only sixty years to double it to three billion.

In the late 1800s a new movement began that convinced some Modernists that some of these new-fangled city ways weren't all they were supposed to be. Their goal was to try and stop progress by idolizing how they imagined life used to be a few decades ago. They believed in mythical small town values. Fundamentalism and racism grew from these movements as people tried to define themselves in purity and simplicity.[37]

The Traditionalist movement was very powerful during this period, introducing such things as prohibition. All over the world the cousins to the American Traditionalist raised their voices. Germany, India, and other ethnic groups began to demand their solidarity in keeping things the way they are, and returning to a better time when their ethnic identity was pure and no other groups were attempting to infiltrate, rule, or contaminate their homelands and cultures. Traditionalism continues on today as a backlash against the influence of others in culture and society.[38]

The twentieth century was the era of the globalization of hatred and murder. Thirty-six million are killed in wars, and one hundred twenty million perished in ethnic cleansing programs.[39] Ethnic cleansing was found as a solution to many a society's problems. We say, "You are eating food that belongs to me." We say, "You threaten me with your weapons of mass destruction so I must kill you and your families first." And through our corporate machines we can even say, "You are living on land I want to use. Move, die, or work for me." We can say this to people around the world through corporations that make deals with corrupt politicians in other nations.

Large corporations manufacture mines, bombs, submarines, aircraft, satellites, warheads, warships, biological weapons, neutron bombs, machine guns, assault rifles, hand grenades, and all the evil concoctions of some dark force while all of it is praised for protecting us, while it's really for imposing the will of corporate exploitation in foreign lands. We stare at evil incarnate and sing patriotic songs of triumph and national anthems that unite the masses in a false sense of belonging. But what we belong to has nothing to do with true humanity and instilling a sense of peace and love and cooperation. It has everything to do with corporate greed.

The Dominator Consumer Economy

The DCE literally began gouging the planet and snuffing out anything in its way as it created a mental environment of pathological desires for wealth and products. To make large short-term financial profits, local and international economies are structured to exploit resources, both labor and natural, to the greatest extent possible, and funnel wealth upwards to the wealthy minority. This linear mode of economics can be briefly outlined to expose its delusion:

1. Corporations locate and obtain natural resources from any place on the planet where they can be procured at the cheapest price,

2. move these raw materials to manufacturing sites for conversion, and then

3. market and sell products to consumers; then

4. consumers use products until products either:

    a. seem to become outdated through new marketing programs to entice consumer base to update, or

    b. become obsolete or break; finally

5. product enters the waste stream and is most likely buried.

This was the beginning of a waste stream unlike any ever known before. Huge waste sites all over the world began to collect and bury billions of tons of discarded trash from corporations and households. This cycle is obviously flawed, because it is not a cycle, it is a one-way road. The price for this type of thinking, of course, will be borne by future generations and will be the legacy of Domination.

The First Generation to Experience a Doubling in the Global Human Population

It took only forty years, from 1960 to 1999, to double the population of the Dominators to six billion. The addition of three billion people to the planet in merely forty years is almost beyond conception, and yet most of us rarely consider it. And now we find that there are barely a few million people left scattered in a few remote spots around the globe that are still considered indigenous, but in extreme danger of assimilation by our global culture. Wild humans are near extinction.

This is the age of corporate power and control. Most of the global economy is now owned by corporations. What few protections remained through anti-trust legislations were essentially eliminated by the policies of the Reagan administration throughout the 1980s.

Repressive regimes are routinely used by corporate interests and propped up on their behalf by the military establishment. The Taliban and Saddam Hussein were once recipients of military aid from the U.S., even though it was well-known how cruelly they mistreated their citizens. The only reason we tolerated their abuses was because we had financial interests through our corporations that were too busy turning profits to upset the status quo.

Altering the Natural World

The landscape of the planet hadn't known such utter fantastic transformation as it experienced in the twentieth century. Areas where swamps, tropical forests, grasslands, lakeshores, valleys, river edges, and other natural places once thrived suddenly turned into shopping malls, suburban neighborhoods, freeways, office buildings, and farm fields. The credo of Progress and the American Way was destroying everything in its path.

A new reality has been built over the former surface of Earth, changing what is available for the human creature to interact with and relate to. Instead of walking on the natural ground, we walk upon grass-seeded turf, pavement, and carpeted floors.

Peter Goin wrote in *Humanature*, "Most Americans still dream that nature is unchanged from what they *think* seventeenth- through nineteenth-century explorers saw in the New World during the era of westward expansion. But we know this is not the case. The air we breathe is an industrial composite . . . Rivers and lakes are elements within a water-management system. Forests are manufactured and harvested like soybeans and corn. Animals are controlled, bred, and genetically designed. Insects are raised in massive numbers, then irradiated and released. Rocks are made 'natural' by spraying cement onto wire forms and adding the right colors. Plants and trees are made from plastic. Beaches are reconstructed. Everywhere I look, nature is an illusion."

Our vehicles separate us from all natural interaction with our planet and other people for many hours a week. Encased in metal boxes we spend many hours every week in transit to get to our destinations. Many hours every day are spent staring at television programming as we ingest obsessive thoughts. Commercial ads and industry propaganda control culture and use psychology to direct our society's behaviors and buying patterns.

Patriotism Displaces the Human Instinct for Tribal Loyalty

In order for tribes to exist in the past, absolute devotion to the tribal unit as a whole was critical. The tribe would not be able to survive if all of its members did not faithfully obey its customs and laws. However, this human trait, to swear complete and unwavering allegiance to an entity of which he is a member and of which is responsible for his and his family's survival, is no longer valid in the new reality.

Tribal loyalty was always natural because each member was an equal part of the tribe to which they belonged. Each individual was the tribe. But a nation-state is an institution created to dominate and impose the will of the ruling class on its subjects.

The origin of a nation's existence was for the purpose of acquiring and retaining agricultural land and valuable real estate. What other purpose could a governmental entity possibly have? In this modern age, many people believe we need our government and the military to protect us from foreign invasion. It is my contention that the German's Nazis, the Communists, the Japanese Empire, the Taliban, and all the purported evil empires and forces in the world are simply various manifestations of our global Domination culture. All the modern nations, no matter what political or socio-economic system label we may give it, are basically totalitarian autocracies. They are all controlled by an elite minority who use the governmental structure, whatever type it may be, for their own personal wealth, benefit, and lust for power.

So, just what is it we are pledging allegiance to anyway? A flag? A governmental system? A landmass? An economic way of doing business? And of course, people the world over are also standing behind their flag. The Italians, the Kenyans, the Australians, the Taiwanese, the Peruvians—all proud to be a citizen of their particular brand of nationalism.

But what's the point? If I'm a Canadian and you're a Russian, I shouldn't look at you as an enemy, but that's what our society teaches us to do—distrust all others. We should distrust other religions, other races, other nationalities, other sexual orientations, other football teams, other schools, other economic systems, other languages, and anything that is other than what we identify with. Our society utilizes a basic human instinct for establishing cultural boundaries, and applies it to our present situation where this instinct has no value whatsoever. Not only is it of no value, it is critically dangerous.

What Really is Sacred?

The Domination culture always seeks to divide, label, and subdue. But for tribal people, the Sacred is found in the connections humans make with each other and with nature. A good friendship or a warm lover is infinitely more spiritual than an angry war-like god that teaches intolerance of others.

I enjoy the timelessness and grandeur of the desert. And when I contemplate the whole of creation in the middle of the desert with the canopy of stars shining down heralding the grand Universe as my senses become overwhelmed by the beauty we are all deprived of in today's society (due to urban light pollution)—to me this is far more spiritual than a dead man hanging on a cross, murdered by others of the same culture because of their own mutual intolerance.

When I walk through one of the remaining forests not yet annihilated by our culture, and feel the ageless life of the planet looming around me, filling my senses with deep and rich odors and lovely plants and trees, I feel more of a spiritual connection to our fantastic blue globe floating in the cosmos than I do in a sterile church built from materials that once were part of the beauty of the forest.

When I sit on the rocks at the shore of the ocean and feel the power of the waves as they crash at my feet, I feel much more of a spiritual connection with the timelessness of all life than I do when I read scriptures about the god of ancient Middle Eastern cults killing populations of people that are the supposed rivals of his Chosen few.

What really is sacred? It depends on your personal and cultural values.

Cultural Collapse

The Domination culture has introduced global war, global famine, economic collapse, ethnic genocide, slavery, crime, and disease to every corner of the Earth. And now that the Dominators have invaded and control the entire world, they've had to project their pioneering myths out into space, as well as deep within, down to the microscopic level with genetic engineering. These are the new pioneers.

The myth they are acting upon, of course, is of the fearless pioneer paving the way for the rest of humanity to our rightful prosperity. So now that we've conquered the planet, the only place we can think of to go is off the planet of which we are a living component. We have many suffering humans in our midst, but choose to send spacecraft into the heavens, as if that could possibly improve the human condition. What is the reason? The myth of the pioneer must be continuously acted upon and repeated over and over.

Almost everyone is at least vaguely cognizant of the multitude of ailments that haunt our civilization no matter how comfortable their immediate surroundings may be. Our vision of humankind's superiority over the planet is crumbling before us. Our institutions become increasingly inadequate of rectifying the multitude of problems that confront us. Our governments and our politicians feed us double-talk. Our corporations keep dumping toxins claiming they are safe. We consume products and create mountains of trash.

People all over the globe are disillusioned and prone to violence and intolerance towards others. Dwindling resources are causing the good people of the Earth to compete with increasing fortitude for that which they deserve—which we all deserve. Clean water, nutritious food, and shelter are things every human being is entitled to. Without these birthrights, people will become more desperate and prone to acts of violence. Revolt, crime, terrorism, and civil uprising are growing and will only become more commonplace as the gap between the rich and the poor become vaster, both globally and locally. Wars will be fought as resources become scarce. Access to oil, water, and food will become issues of the highest priority not only for individuals, but for nations and cities. We will fight for these things.

Then there are those who don't have the energy or capacity to wage a fight to the death for survival. For them, drugs and complacency shape their existence. In fact, the antidepressant market soared by 16 percent every year between 1989 and 1999 in the G7 nations.[40] For these people, our cherished myths are increasingly proven to be meaningless in today's stark realities. More people are resigning themselves to a hopeless cycle of repetitive mundane actions, totally at the mercy of the unseen powers that be, while witnessing what seems to be an unstoppable decline in life quality from the personal to the global.

Why has our species become the enemy of life? Our hate and contempt ranges from tiny bugs and weeds to entire subsets of our own kind. What is the source of our cultural anger? Our desire to dominate is as invisible to us as it is vast. It is buried deep in the subconscious realms of our conception of reality, from any one individual to the core of our culture's belief system. We even think that such sickening and evil desires to dominate are fundamental qualities of our species. More and more of us are reaching the conclusion that something very deep in the roots, the very foundation of our culture, is what is destroying our cherished mother of all life, the Earth, and there doesn't seem to be anything we can do about it.

The values and spiritual connection to the planet and each other that we humans evolved with have become preempted with greed and empty slogans that appeal to the lowest common denominator. People struggling with the daily chores of survival have become disenchanted and uninformed in a sea of trivial and useless information that distracts their attention from the corporate concentration of financial wealth and power. Individual ideals have been broken and compromised by cheap, vicarious, and meaningless experiences that make life a parody of what it is to be human.

What we are experiencing is the collapse of our culture. It is not unprecedented; many cultures have collapsed throughout the course of human history. Of course, when they did there were still thousands of other cultures that survived.

Diversification vs. Homogenization

We are slowly dissolving all diversity on many levels. The approximate number of autonomous political units in 1000 BCE was around half a million distinct bands, tribes, villages, and chiefdoms. Fifteen hundred years later, by the year 500 CE, that number had fallen to two hundred thousand. Today, that number has fallen to barely two hundred units the world over.[41]

But on a cultural level it is even more homogenous and less diverse as we now have primarily just one global culture. From Japan to Australia to Africa to Mexico, we're all pretty much the same. The differences between the subcultures seem huge to us, but we do not as readily see the Sameness that towers so large, and is so pervasive that it is all but invisible to us. This is because we take as a given those aspects of each subculture of the Domination culture which are of a similar nature. Although we may think the planet has many cultures, it really is primarily just one huge global culture (with many familiar and readily identifiable subcultures). While we may be quite able to point out the differences between the people in one part of the globe from another, we are much less capable of realizing the similarities. This is because the similarities are viewed as being traits we might assign to the human condition, and not an aspect peculiar to our global culture.

Because the global culture is so incredibly expansive and pervasive, we think of ourselves as humanity itself.[42]

For example, if we can see that people who live in Canada and Italy and Argentina drive automobiles, have corporate jobs, attend Christian churches, pay rent, and purchase production line corporate food at the grocery store, we can only imagine that this is how it should be. This type of behavior appears to us as a "human" phenomenon. It seems to us a universal Truth that this is how humans should live. We take it as if it always was, and always will be. We see it as a self-evident truism.

One World—One Culture

It is just one global mega-culture now that is responsible for determining the fate of all humanity. The lack of diversity between the ethnic subcultures is becoming even more prevalent as the corporate machine dominates the flow of information throughout the world in its attempt to create large predictable markets for various products. People are primarily being exposed to a lesser variety of viewpoints. As the evolutionarily sound concept of diversity is undermined, all of human culture and beings are becoming more and more alike.

The Bottom Line: *If the Domination culture fails, our entire species is vulnerable*, not just one or two tribes. And since the Domination culture has also taken control of natural systems, such as streams, the surface of the land, the chemical composition of the water and air, and the genetic makeup of both plant and animal species, these entire systems are also vulnerable.

Deceived

The Dominator attitude holds that the Earth is but some rude way station in the human journey towards perfection. Earth to us is like some sort of lowly purgatory. Early agrarian people hated the toiling they had to suffer in order to survive, so perfection was imagined in death. To Dominators, Nature and all of the world and its creatures is lowly and crude. Deep inside our imaginations is the idea that God is coming to witness the inevitable demise of humanity in the Armageddon. The Earth will be destroyed as the faithful rise to everlasting life in heaven. We live as if our actions are merely moot points in humanity's predestined demise and hopeful salvation. We live as if this story line that began with Adam and Eve is the actual story of all humankind.

As the progenitors of our present day religion, our cultural ancestors have placed our current generation in a precarious situation. We worship an alien (God) that speaks to us of a better place (Heaven) than what we have right now. We imagine we are here to prove our love to a deity we'll only meet when we die and that the Earth is but a temporary testing ground.

The deception is obvious. How can we manage something our culture detests? The Earth's ecosystem is not a machine that our experts can run. The totalitarian agriculturists were wrong. We evolved to exist within the laws of nature, not outside them.

With the deception exposed, we need to further examine the measurable effects of this culture before we can start to convince others of the need for change. Indeed, it was these facts, which I will be presenting in the next two sections, that created this overwhelming desire in my own consciousness to want to transform the unchangeable.

II

Effects of Domination

○ ○

"On the altar of short-term profit, before the false god of consumerism, we are plundering the world, putting our children's future at risk, and even most educated people don't realize how or why it's happening."

—*Thom Hartmann*

Change the Subject

Of course, who can really find pleasure in thinking and reading about the myriad of travesties that are destroying our world? Well, this section covers just that and only that, and because of the extensive nature of degradation that I've uncovered, this is also the longest section of this book. Because this section is not concerned with the solutions, the situation can begin to appear so mired in a seemingly endless tirade of hopeless facts and feedback loops that continually overlap and amplify that it seems useless to go on! Please, don't despair. There are solutions, and they will follow.

Society's Misplaced Priorities

When the media covers environmental degradation issues, many times their focus is on the effect these problems are having on our society and not what effect our culture is having on the environment. I once read an article about the depletion of salmon in certain rivers in Alaska. This article held my attention as I read with great interest, expecting to learn the possible explanations for this demise. I was astounded that not one theory was advanced as to what the possible causes of this seemingly catastrophic situation could be. The entire article focused on the profits and losses of the fishing industry in Washington.

The intervention of our highly industrialized, corporate, consumer-based society on the natural world will come back in many untold ways we've yet to learn. We can all feel sympathy for those of us who depend on skills acquired over a lifetime of hard work and dedication who find themselves unable to provide for their families with those skills. It is true that our social system, our society, reality as we presently perceive it, is wholly and thoroughly dependent on our economic system. But what we, as a culture, cannot accept is that our very existence is wholly and thoroughly dependent upon the planet's ecosystem. This is irrefutable fact. It is undeniable truth.

What we as a society, as a culture, and as individuals need to come to fully understand is that the very fabric of any human society will disintegrate if the ecosystem upon which it depends is destroyed. There is just no way to circumvent this truism.

Oil

The infrastructures of most all the world's societies are designed to be powered by oil. In countries where oil is the primary source of revenue, bad things usually happen to the quality of life for the people who live there. The very nature of the industry requires a centralized power structure that seems to be perfect for dictators. Dictators work quite well with corporations that need cooperation from the local governing authorities in order to be able to exploit the environment and the local population to create the highest possible profit margin. This requires force and intimidation. The western world will always encourage strong centralized governments and dictatorships in countries with large oil supplies. They might complain about their harsh methods, but are secretly appreciative of their control and usually eager to do business with them.

Iraq, Saudi Arabia, and other such governments have harsh regimes that utilize cruel methods of controlling their people. Nigeria is a prime example. The Ogoni tribe of Nigeria is a small minority that has a rich oilfield under their homeland. Yet these people are impoverished in what was once a rich diverse ecosystem. Forced into poverty, they have no future due to the complete degradation of their own land by a few oil companies who provide the income that funds the brutal dictatorship, which then has the means to subdue the local population.

Aaron Sachs writes in *World Watch* magazine that "unlined toxic waste pits allow pollution to seep into drinking water; open gas flares destroy plant life, cause acid rain, and deposit soot on nearby Ogoni homes; and corroded pipelines crisscross the Ogoni's fertile agricultural land, rendering it economically useless . . . some tributaries of the Niger run black."[43]

Oil spills are a regular feature of this industry. The Trans-Alaska Pipeline has had many barrels spilled on pristine areas of Alaska.[44] The potential for complete unmitigated disaster looms over this travesty of environmental crime, putting rivers and species of all kinds at grave risk. Everyone knows about the Exxon Valdez oil spill, but this happens all the time and barely makes the news. It seems every year I hear about another oil tanker disaster somewhere in the world. Of course, it hits the news and soon fades from public attention as the lingering effects of each incident is rarely followed up on by the global monolithic news services.

Unfortunately, this is a rapidly escalating situation. The oil and gas industry annually spends $150 billion looking for new oil reserves. The threat this foists upon old growth forests, coral reefs, mangrove ecosystems, and indigenous populations[45] is unforgivable and a tragedy of the most immense proportions.

Corporate culture has created an infrastructure that is based upon oil, which is dangerous because it pollutes and creates an exploitive relationship with human culture, civic structure, and political power—not to mention the fact that everyone knows, although the exploitive institutions would never admit it, that there is only a limited supply of oil left and that at present rates of depletion and demand (which is expected to rise) we will run out of oil in the next half century.

Of course, massive struggle for the rights of oil (which gets increasingly expensive to extract as time moves forward) would trigger civil and global strife unlike any ever known—and this, too, is being masked by those who would lust for profits and power, though the telltale signs of a possible future war and loathing between the dominant global factions and the oppressed grow ever more visible as they are being increasingly manifested in our news and consciousnesses.

Mining the Planet

Mining corporations around the world gut the planet to extract various minerals. When they finish the annihilation of one area, most of them walk away leaving the public to clean the toxins. Cyanide and other chemicals that are used to separate the ore seep into wells and aquifers.[46] This is our legacy of the environmentally destructive indifference we are leaving for future generations. Migratory birds that drink this toxic brew die by the thousands. Thousands of miles of once clean rivers and tens of thousands of acres of lakes have been polluted and spoiled by this industry, leaving them unable to support life.

Archaic U.S. laws allow foreign companies to claim U.S. land and extract billions upon billions of dollars worth of precious minerals for just a few thousand bucks. Around $3 billion is taken from U.S. public lands annually. There is a mine in Nevada that contains more than $10 billion in gold. The price paid to taxpayers by the Canadian corporation that is excavating this precious mineral—$5,190.[47] When one considers that 85 percent of the gold mined in the U.S. is for jewelry,[48] maybe that is one luxury our burgeoning numbers could forgo in return for a cleaner environment.

Greedy corporations the world over are extracting precious metals from deep inside the planet, and filling their own coffers with cash at the expense of the environment and the indigenous peoples who live near the mines. There is a mine located in Indonesia that extracts 165,000 tons of ore every day from a mountain, and then dumps 98 percent of that in a river. It is contaminating downstream forests with arsenic, lead, and mercury among other poisonous wastes. In the process, the operators of the mine kill fish and destroy the water supplies for the local population, not to mention the only way of life that has sustained them for thousands of years.[49] Over its lifetime, the mine is expected to dump over 3.2 billion tons of acid-generating waste rock into the local river system.[50] This industry is killing the life in this river, as well as hundreds of others worldwide.

Deforestation

Currently, less than 5 percent of American and European native ancient forests remain.[51] One third of the forests in the entire continent of Asia have been lost since 1960. The planet is losing, through clear cutting and burning (primarily for meat production), 2.4 acres of virgin rainforest every second.[52] The second-growth forests that have been allowed to grow back are not as genetically diverse and are more vulnerable to disease, fire, and pests than their wild counterparts.

In the 1800s, more than 160 billion board feet of white pine were harvested in Michigan, Wisconsin and Minnesota. This brought the deforestation of an area the size of Europe. And since 1955, the ancient forests of the Pacific Northwest have been destroyed along with all the natural evolving life that lived in these areas.[53]

Lumbering campaigns sponsored by the industry tell the public that salvage logging assists the natural forest by "harvesting" only the dead or dying trees. Forests have existed for billions of years replete with dead and dying trees and thrived. These dead and dying trees are critical to maintaining a healthy forest.

They absorb water and retain it. Their downed trunks provide a slow release of nutrients into the forest floor and are home for countless diverse species of life.[54]

Lumbering advocates are always quick to point out the jobs that this horrible industry provides, as if the local economy was more important than the local ecosystem. They also claim that they replant the forests they cut down, making the resource available to future generations.

What they are not telling us is that monoculture tree farms that take the place of ancient ecosystems support life about as well as a crop of corn. Plus, they are much more vulnerable to disease because of the lack of diversity. One aggressive species of pests can destroy an entire forest because all the trees are identical. A genetically identical "forest" creates a cornucopia of food for opportunistic pests who feast upon a large tract of man-made abundance and sustenance. Even when they are mature (before they are harvested once again) they only support a limited number of wildlife because certain species will not inhabit forests that only contain one type of tree. This has the effect of driving a vast number of symbiotically related species into abject extinction because they have no place to inhabit. The bottom line is that a tree farm will never take the place of a diverse and rich ecosystem in the web of life. Planted monoculture trees cannot be called a forest—it is a farm.

A very important function of trees is to stabilize and hold the soil together to prevent the ground from disintegrating when it gets wet. When we alter our forests by ripping them apart, there is more lost than just the trees themselves and the ecosystem they support. The rain rushes down the valleys and instead of soaking into the lush forest floor, floods the rivers with thickened muddy water. It carries away the precious topsoil that holds the hope of nourishment for future generations. Seventy-one percent of landslides are directly attributable to fresh cuts, with 23 percent in older cuts.[55] Only 6 percent occur in land that has not been cut. Trees produce oxygen, purify water, maintain biological diversity, regulate the chemical composition of the oceans, produce genetic and medicinal raw products, and maintain wildlife habitat and migration patterns.[56] Snuffing out life on Earth through mowing down forests causes floods, mudslides, forest fires, fouled water,[57] and a dead planet.

In order to get to the resources that are hidden deep within remote areas, roads have to be built. In the U.S. alone, the Forest Service has constructed more than 360,000 miles of them. That's eight times the length of the Interstate Highway System. These roads pollute streams and forests.[58] They also open up sensitive wildlife habitats for commercial exploitation that would otherwise be inaccessible.

The Landscape and Land (mis)Use

We have altered most of the planet. Most of what we see today was not there two hundred years ago. Farms, cities, malls, factories, freeways, and suburbs are sprawled over land upon which once lived forests, marshes, wetlands, and grasslands. Now many are vanquished from the face of the planet. Forever. We have paved the natural world and created a new reality. Yes, it is still the same planet, but it is altered now beyond recognition.

Even out on the so-called "lone prairie" where pavement and buildings are rare, you couldn't walk very far without encountering a barbed-wire fence. These abominations of the natural order crisscross the entire land surface of the planet with no regard to natural cycles of the wildlife that once ran free. Now mammals and birds by the thousands become entangled and die every year in these unholy divisions that halt migration.

Not only are we permanently scarring the face of our planet, we route our stinky, messy, toxic garbage back into its surface, as if no one will be bothered by it there. Americans waste and send more trash to the dumps than any other nation, per capita and as a country.

Carpets, plastics, batteries, paper products, furniture, metals, disposable diapers, disposable cameras, broken kitchen appliances, organic matter, chemicals, and much more are dumped into our environment every year. Year after year, the production of chemicals and the extraction of gas, coal, oil, and minerals generate hazardous wastes. In the U.S., 85 percent of trash gets buried in toxic tombs we call "landfills."[59] Mountains of non-degradable spent consumer packaging and products will remain in these landfills of our Earth as we bargain for more precious land to stash our trash.

Many landfills are now found to contain much more than just table scraps. Toxic household and industrial chemicals are also being buried and are finding their way into our air and water supply. As precipitation drains through a dump, it carries with it various contaminants down into the ground water.[60]

The National Safety Council is predicting that as many as 680 million computers are expected to enter the waste stream before 2010. This stream would include:

- 4 billion pounds of plastic,
- 1 billion pounds of lead,
- 1.9 million pounds of cadmium,

- 1.2 million pounds of chromium, and

- 400,000 pounds of mercury.

Most of this electronic waste gets "recycled," euphemistically speaking, to Asia where certain villages have come to depend on the computer recycling market. Parts that are not resold get buried locally, allowing toxic components to leech into the groundwater causing higher rates of cancer, neurological, and other degenerative diseases.

Cell phones by the hundreds of millions are slated to enter the global waste stream soon. Of course, the same toxic materials found in computers are also to be found in these indispensable yet disposable accessories for the twenty-first century human—arsenic, antimony, beryllium, cadmium, copper, lead, nickel, and zinc.

We like to think that the stuff we toss gets sealed away in some distant repository, never to return. Guess what? It's still here—right here on Earth (a closed system) where we live. And it's not only the garbage that you personally tossed out last year and the year before that, it's all of your neighbors' garbage too. Then throw in the even larger amount of wastes from all the businesses and manufacturing facilities in the area. Added together, it equals a travesty on our home that is entirely preventable.

Furthermore, a huge portion of the planet's surface is dedicated to the production of meat. Half of our planet's land surface is dedicated for grazing animals. Two-thirds of the farmland in the U.S. goes not for food for humans to eat, but feed for corporate-owned animals. Similarly, two-thirds of U.S. agricultural exports are feed for animals.[61] Eighty percent of the corn and 95 percent of the oats grown in the U.S. is fed not to humans, but ravenous big fat cows.[62] We stuff these hapless animals, while our own fellow human beings starve to death by the thousands every day. Annually, U.S. cattle, pigs, poultry, and fish devour 160 million tons of feed grains. It's actually quite amazing when you think about it; 1.2 BILLION acres of land in the U.S. alone is dedicated to growing *feed* instead of *food*.[63] (How much is that? 1.2 billion acres is equal to the combined land area of Texas, California, Montana, New Mexico, Arizona, Nevada, Colorado, Wyoming, North Dakota, South Dakota, Pennsylvania, New York, North Carolina, South Carolina, Florida, Georgia, Illinois, Wisconsin, Indiana, Kentucky, Tennessee, Virginia, and Maine.[64])

Most people believe that the only reason forests are destroyed is for lumber and paper products, but this couldn't be further from the truth. As the demand for meat skyrockets by a burgeoning and increasingly affluent population eager to climb up the food chain, more crops for feed and pastures for grazing become necessary. This drives up land values, and land-hungry corporations twist the governments' arms or bribe their politicians to gain access to public lands at public expense. Forests fall forever as cows and other livestock replace them.

Mangrove forests (coastal swamplands) that grow throughout the tropics are being decimated for shrimp farms to feed the wealthy. These trees have tremendous ecological value that is not considered in the race to make up for the dwindling wild oceanic harvest. As these critical systems are torn apart, the safe nurseries for thousands of other species is lost and they head for extinction.[65] Mangrove forests also buffer the land against hurricanes, filter out toxins before they get into the ocean, and provide safe habitats for a wide variety of creatures, from alligators to monkeys.[66]

Golf courses are another grotesque insult to our planet. These huge swathes of former diverse ecosystems are poisoned by managers who use more insecticides than do farmers.[67] Golf courses can apply about 50,000 pounds of chemicals on the courses, which translates into eighteen pounds per acre, which is about seven times as much as commercial agriculture.[68] Golf courses consume up to 800,000 gallons of water per day.[69] With the pollution of the Earth's environment and depletion of our valuable resources, we need a global outcry to halt the expansion of this ecologically destructive sport. Considering that, on the average, a new golf course is added somewhere on the planet every day,[70] there is no time to waste.

We don't often stop to consider that our homes, workplaces, and farmlands used to be a diverse ecosystem. But as our population skyrockets, residential and commercial developments grow and overtake forests and farms. As vast tracts of fertile soils are destroyed with suburban sprawl, where will the food come from to feed that population?

Soil Erosion

Life on the land surface of our planet emanates from the soils. Any high school student knows how a forest ecosystem continually replenishes and recycles life. We are taking crops and trees created by the soils. The problem is that as the forests vanish, so do the soils underneath which are washed away by rainwater or are blown away by wind. If you remove the trees and bare the soil, in a few decades those soils become depleted of life. It is only how deep the soil is that determines how long that process takes. It must and will happen because there is nothing left to renew the soil.

Cattle grazing is by far the primary cause of the spreading desertification that is destroying life as we know it.[71] Thirty percent of the land mass in the U.S. is currently used for grazing, and 30 percent of that is land that is publicly held and leased out to corporations on the western range at bargain basement prices.[72] Industrialized livestock production is quite literally eroding the soils upon which all land life, including human life, depend upon for survival. And the tab for all of this devastation is picked up by the taxpayers who support this industry by installing water pipelines, building fences, providing cattle guards, seeding the lands, weeding out plants that are not conducive to the industry, and destroying wild animal populations that are deemed to be harmful to domesticated livestock.[73] Taxpayers actually end up paying this industry over $1 billion annually in the direct and indirect costs mentioned above.[74]

When fertile land is stripped, it eventually turns into desert. The Sahara Desert has been moving southward at the astonishing rate of 30 miles per year.[75] In Africa, the soils are becoming barren and unable to support life. Yet, just one hundred years ago it was teeming with wildlife and diverse vegetation. The rangeland for cattle in Africa grows by millions of acres every year.[76] Areas with access to water raise more livestock and are trampled, stripping bare the vegetation.[77] Eroded and stripped from overgrazing, it lies barren and impoverished, unable to support animals and the people who depended on it for their survival and livelihood.[78] Is it any wonder why the people who live there are having a problem finding food? As millions of rural refugees try to escape the growing area of desertified land,[79] we can see what might be in store for us someday.

This is not some distant planet that has no effect on us. Indeed, we are the neighbors to this growing malady. Most of us have yet to comprehend the magnitude of this phenomenon, what the implications are for the future, and what possible effect this could have on us.

Streams and Lakes, Our Life-Sustaining Waters

Hordes of cows crowded together can be found just about everywhere down the backcountry roads of the American west.[80] The whole countryside stinks, as bovine waste finds its way into streams and rivers. These sorry looking, genetically-bred-to-be-big-and-fat creatures now infest the canyons, plains, hillsides, meadows, river banks, and lake shores of the entire planet, trampling, chewing, and literally dumping on the rest of creation.

Their wastes are not treated in sewage plants like human wastes; it is dumped onto our soils where it blows away in the wind or seeps into our water supply.

U.S. industrial waste is a scourge on our fresh water supply, but few realize that animal industries account for twice the amount of organic pollutants.[81] One steer produces forty-seven pounds of waste every day.[82] All farm animals produce about ten times as much excrement as the entire human population.[83] Collectively, in just our country alone, cattle produce 230,000 pounds of manure every second, 1.6 million tons annually.[84] These wastes create ammonia and nitrates that pollute our wells, rivers, and lakes, contaminating our drinking water,[85] and killing the life in these once pristine waters. Eighty-three percent of the streams in the state of Wyoming have been lost.[86] It is the unnaturally excessive nitrogen found in manure that creates these toxic compounds. The web of life here will sooner or later be missed once we realize the ramifications of carelessly allowing huge populations of cattle to be concentrated near the life bearing waters of our planet.

The average corporate hog farm produces more waste than a city of 100,000 people. Waste from these pig farms breed bacteria and viruses that enter the human food supply. These wastes are so toxic to our waterways that these farms kill thousands of migrating waterfowl every year.

There are around fifteen thousand publicly owned sewage plants in the U.S. that collectively produce about eight million tons of sludge each year. This sludge consists mainly of human excreta, but also contains biologic pathogens, heavy metals, and chemicals.[87] Half of this sludge is applied to farmlands where it can leach into streams and into the ground.

In their book, *Blue Gold*, Maude Barlow and Tony Clarke write, "Half the people on this planet lack basic sanitation services. Every time they take a drink of water, they are ingesting . . . water-borne killers . . . Eight percent of all disease in the poor countries of the South is spread by consuming unsafe water. The statistics are sobering: 90 percent of the Third World's wastewater is still discharged untreated into local rivers and streams; water-borne pathogens and pollution kill 25 million people every year; every eight seconds, a child dies from drinking contaminated water; and every year, diarrhea kills nearly three million children . . ."[88]

Fresh water sources are rapidly being polluted beyond repair from medical wastes, chemicals, pesticides, nitrates, phosphates, and radioactive wastes that are leaking from industrial sites the world over. Gasoline tanks, sewage lagoons, landfills, feedlot effluent, mine tailings, oil spills, and road salt are utterly devastating groundwaters and aquifers. Some of these substances are lighter or heavier than water so they float on top or sink to the bottom of rivers, aquifers, and lakes.[89] These pollutants promise to alter and harm life for many millions of years.

And as life in the streams decline, we stock them with hatchery hybrid fish to make popular fishing spots appear healthy, even though they are dying. These inbreeds learn incorrect feeding strategies before release that are not viable in a wild habitat. Most are caught, few reproduce, and they tend to be highly competitive and fight with the wild population. For example, more than two hundred species of fish disappeared from Lake Victoria in Africa in the 1980s and 1990s due to the introduction of a voracious alien fish called the Nile perch.[90]

Fish populations are dropping fast, and few really know the extent of it. For example, over forty-two million pounds of salmon and steelhead were caught in the Columbia River in 1884. In 1994, that number had dwindled to only 1.2 million. The Rhine River once was heavily populated with salmon, but by 1950 they were gone.[91] Over-fishing, bio-invasions, and chemical pollution have all but wiped out the native salmon and trout in the Great Lakes.[92] This story is being repeated the world over.

There are estimated to be around forty thousand large dams in the world today, flooding an area as large as the state of California and displacing and uprooting over ninety million people. Dams are a curse on life in their vicinity as the submerged vegetation decomposes and emits vast quantities of carbon dioxide and methane. Mercury is also released from former soils as it becomes bioavailable to algae, and consequently to fish and the entire food chain where it accumulates in the fatty tissues of all creatures. Excessive evaporation due to an increase in surface area raises the salinity of the rivers. The salt then destroys aquatic life, wetlands, and surrounding soils. Riverbeds and water channels are totally snuffed out as they are buried in sediment, causing entire ecosystems to collapse as symbiotic relationships between species are lost.[93] Additionally, whole communities and cultures that formerly inhabited flooded areas are destroyed.

Many of the planet's great rivers, once they navigate through the farmlands, dams, and municipal water supplies, are reduced to a toxic trickle by the time they reach the oceans. The Colorado River used to pour into the Gulf of California. It doesn't even reach the sea anymore. The Yellow River in China doesn't make the trip to the ocean for most of the year. In Africa, the Nile has very little water left by the time it gets to the ocean.[94] Droughts and overuse of water supplies are causing our natural sources of water to become depleted and polluted. Paint thinners, grease, oil, and other toxic industrial wastes are fouling water supplies in every corner of the world causing massive die-offs of species that are dependent upon these resources for food and shelter. Lakes are drying up and rivers are running dry before they reach their destinations. Billions of people will be faced with vast shortages of available water in the coming years if we continue on this path.[95]

Few people realize that the High Plains states of the U.S. (Wyoming, South Dakota, Nebraska, Colorado, Kansas, Oklahoma, Texas, and New Mexico) were not suitable for farming until the Ogallala Aquifer, the largest underground reserve of freshwater in the world, was discovered in the 1800s. Now thirteen trillion gallons of water are pumped from the Ogallala Aquifer every year, primarily for beef production. Irrigated crops provide feed for livestock. The water from the Ogallala is essential for the livestock production and meat processing industries. The crop, livestock, and meat processing sectors form the core of the economy of these states, accounting for a large share of employment and gross output.

But the Ogallala recharges very slowly; the High Plains economy is dependent on a finite resource. If we continue on this way, it is projected to run dry in less than thirty-five years. If that should occur, the economy of this vast portion of the U.S. would surely collapse, not to mention be unable to sustain human life.

We can look forward to many conflicts between nations and various industries for fresh sources of water, and a corresponding rise in the cost of water and the products that depend on it. Residential water use is skyrocketing, but less water and land for crops and irrigation will increase the demand for importing food from other locales that will soon be experiencing the same dilemma. Rapidly decreasing availability of food and water and increasing social unrest will occur in this inescapable situation. Free Trade agreements such as NAFTA are now resulting in privatization of water *rights*, and the public loses control over the most basic human right. As profits are extracted, the cost of water rises and the poorest people are priced out of the market. With half of the world's population without reliable drinking water and access to sanitation,[96] disputes and conflicts are breaking out around the world.

The Systematic Elimination of Oceanic Wildlife

The species of fish currently harvested are much lower on the marine food chain than just a decade ago. This loss of biodiversity should be a warning that we are exhausting the life of our oceans.[97] All of the world's major fishing areas have reached their natural fish limits and are in decline.[98] That is not stopping the fishing industry. In fact, the demand for seafood is rising. Currently only about two-thirds of the fish caught are for human consumption. The remaining third are ground up and fed to livestock.[99]

Corporate-owned factory trawlers purchased with low-interest government loans rape the seas with giant nets that snare all forms of life indiscriminately. Those animals trapped in this nylon web of death that are not edible are ground up and regurgitated out the rear of these floating factories. As they perform their gruesome task of purging all life in their path, countless species are driven to extinction due to outright murder or denying other predator species their food source. Fish-eating seabirds and marine animals starve to death because their food supply is gone.[100] Trawling also scrapes the bottom of the oceans clean, killing the life forms that live there.[101]

Every pound of shrimp also brings about the death of up to twenty pounds of other sea creatures.[102] Dolphins, turtles, seabirds, swordfish, sharks, jellyfish, and juvenile fish are among the many "innocent bystanders" that are killed by this industry.[103] Shrimp trawlers produce less than 2 percent of the world's seafood, yet waste a third of the total oceanic bycatch as they drown up to 150,000 sea turtles annually.[104]

Poisoning the Oceans

Oil spills and industrial and household wastes are causing massive fish die-offs. Factory farms leak tons of manure, agricultural chemicals and fertilizers into streams and rivers, which wash out into the oceans. The San Francisco Bay is polluted with such toxic pollutants as methyl mercury, polychlorinated byphenyls, dioxins, and pesticides.[105]

Pig farms in North Carolina are polluting the coastal waters, providing ideal breeding conditions for a microscopic organism called pfiesteria. This cell contains a powerful nerve poison that kills fish and can infect humans. As these farms multiply in numbers and increase in size, more pathogenic organisms can be the only result. Huge lagoons full of decomposing waste are bursting at the seams and leaking into our water supply. It is estimated that pfiesteria has wiped out close to one billion fish in North Carolina.[106]

In the Gulf of Mexico there is an area covering over seven thousand square miles that is devoid of all life. It is called the Dead Zone and it is growing bigger every day.[107] Do we need any further indications of the direction we are headed?

Air Pollution

The ozone layer protects the life on our planet from the ultraviolet radiation that is continually bombarding us from outer space. This precious resource is such a gift to all life, I find it amazing that its value is not held in higher regard. Without this thin fragile layer, life would be quite different. Plants, animals, and delicately balanced ecosystems would perish. But scientists have now learned that we are depleting this layer. Every year the hole in the ozone layer over Antarctica has been getting bigger and bigger. In 1995, it reached twenty-four million square kilometers.[108] Now we're learning that it's getting taller. The zone of complete ozone destruction grew from 8.7 miles tall to 12.7 miles in 1997.[109] We can expect that ozone loss will increase over the coming years and the levels of ultraviolet radiation will likewise become increasingly prevalent in the northern hemisphere, becoming particularly heavy in the spring.[110]

Jet aircraft spew excessive amounts of soot and sulfuric acid, and are the only source of pollutants in the stratosphere. They also pump nitrogen oxide and carbon dioxide (among many other toxic emissions) into the troposphere creating ozone clouds that trap heat, as daytime contrails (the trail of condensed moisture behind jet aircraft) reflect heat back into space. This phenomenon can change weather patterns as temperatures in the atmosphere can alter the intensity and frequency of storms. Contrails also add vast amounts of water vapor, creating more than the normal amount of cirrus (high altitude) clouds.[111]

The level of carbon dioxide in the atmosphere has increased by 30 percent over the last fifty years![112] In fact, the concentration has not been exceeded during the past 420,000 years, and perhaps as much as the past twenty million years![113] Up to a quarter of this gas is the result of deforestation.[114] Up to 20 percent of the world's carbon dioxide emission comes from cars and trucks.[115] By 1990, there were over 500 million automobiles in operation,[116] and that number is expected to double to one billion by 2030. Citizens of the industrialized nations, such as the U.S., are driving the most fuel inefficient automobiles as well as driving these gas hogs greater distances with each passing year. On the average, every car that is used on a regular basis annually emits five tons of carbon dioxide.[117] Urban areas are heavily polluted with another carbon-oxygen combo due to motor vehicle exhaust—carbon monoxide. Carbon monoxide is a colorless and odorless poison that infiltrates our blood and saps it of oxygen.[118] Carbon emissions from the burning of fossil fuels is picking up speed in the developing countries, more than doubling them over the past decade.

The production of cement produces 7 percent of the carbon dioxide that our industrial society emits into the atmosphere. This percentage will change as cement production is growing annually by 5 percent.[119]

Disposable products that are made with paper, such as napkins, toilet paper, and tissue, add vast amounts of carbon dioxide to our air when they are burned, and methane gas when they are buried in landfills. It is estimated that the world's landfills account for up to 19 percent of the global methane emission.[120]

Large-scale agricultural practices also adversely affect the atmosphere. Nitrogen is released into the atmosphere from commercial fertilizers. This alters the Earth's climate by eating away at the ozone layer as well as the creation of smog.

Deforestation has caused an exponential rise in the world's population of termites—one-half ton per human! Termites, unfortunately, generate vast amounts of methane gas. Large amounts of methane in the atmosphere are also produced by the hundreds of millions of cows we raise for hamburgers.[121]

Sulfur is also being pumped into the atmosphere as a by-product of the burning of fossil fuels and from metal smelting, oil refining, and the manufacturing of paper.

Sulfur dioxide and nitrogen oxides cause acid rain which destroy crops, forests, and aquatic life. Together with carbon dioxide, methane, and a host of other airborne pollutants, these levels are on the rise. These substances are creating the greenhouse effect, which is raising the temperature of the atmosphere at the surface of the Earth. This is known as Global Warming.

The United States produces 25 percent of the greenhouse gasses. This is a staggering amount when one considers that the U.S. only accounts for 4 percent of the world's population, not including what U.S. corporations are emitting in other parts of the world on our behalf. It seems quite obvious that the United States of America is the only country that can turn this situation around because we are far and away the largest polluters per capita.

This Stinks

I'll never forget driving through certain industrialized sectors just south of Chicago where it would seem that life itself was nonexistent. The landscape is proliferated with rusted warehouses and billboards as plumes of soot and multicolored smoke rise toward the sky. The stench of fumes and chemicals choke the very life out of the air the residents breathe. I've also seen similar areas around Los Angeles and San Jose, California. I'm sure that just about everyone has witnessed something like this; they're all over the world. Natural life is not evident in these places, and I have a difficult time accepting that this could be a good thing. How is it that this is allowed to happen?

Most of us have probably experienced at one time or another walking in a downtown district of a large city, finding it hard to breathe because the fumes in the air were loaded with toxic chemicals from engine exhaust. Even in rural areas most people have experienced the air thick with the odor of cow excrement from intensive cattle production. How can a person feel that life is good when they can barely inhale? Is this quality living?

Altering the Atmosphere Evokes Changing Weather Patterns

The gasses we've just covered, along with many others, are already measurably altering the temperature and changing the weather patterns beyond our mortal control. The global-average surface temperature has risen by 0.6°C over the 1900s.[122] All fourteen of the warmest years since 1860 have occurred in the last twenty.[123] July of 1998 was the hottest month in earth's recorded history,[124] and the fifteenth consecutive month that the average global temperature set a new high record for that particular month.[125] As the weather patterns are permanently altered, so is life itself, immersing Earth's ecosystems in a profound crisis.

The Earth's temperature rose two degrees Fahrenheit over the past one hundred years and is still on the increase. This warming phenomenon is melting the ice mass of the North and South poles. Alaska's average surface temperatures have increased by 5°F, and this increase is responsible for melting glaciers, thawing tundra, and sinking and dying forests.[126] The North Pole melted for the first time in human history in the summer of 2000. In the late summer of 2003 the Ward Hunt Ice Shelf on the north coast of Ellesmere Island fragmented into two main parts and several smaller ice floes. As this happened a fresh water lake emptied into the ocean waters. And nearly 90 percent of the ice shelf along the northernmost landmass of North America has disappeared.

Warming the atmosphere also causes increased levels of methane and carbon dioxide to be added to the atmosphere when these gasses, trapped in polar and glacial ice, are released as these massive glaciers melt. This triggers a feedback loop by accelerating global warming causing the release of even more greenhouse gasses.

The average temperature has increased by 2.5°C since the mid-1940s at the Antarctica Peninsula. Temperatures there are rising five times faster than the global average.[127] As a result, huge portions of ice, miles thick, are vulnerable to breaking apart. Two large sections known as the Larsen A and B ice shelves are crumbling into the ocean in sheets as large as 1,300 square kilometers.[128] The Larsen B ice shelf currently covers 8,000 square miles. Even larger sheets of polar ice could soon break apart. The phenomenon we are currently observing at Antarctica does not bode well for the future of life as we know it. If ice packs should break apart, which is entirely possible, it would throw the planet into the most horrific of convolutions. As has happened in the past, about 120,000 years ago, the sea level could increase by twenty feet (when ice packs melt), and then drop by fifty feet (as an "ice age" is precipitated), all in the short space of one hundred years.[129]

Other frozen places are melting too. Non-polar glaciers have retreated all over the world during the 1900s. The extent of spring and summer sea-ice has decreased by up to 15 percent since the 1950s. The thickness of Artic sea-ice has declined by 40 percent during late summer to early autumn.[130]

The temperature of the sea surface off the coast of California has risen by as much as 1.5°C since 1951.[131] Climate change is harming microscopic phytoplankton plants due to sea surface temperatures, which have risen about 3°F over the past forty-two years. These plants form the foundation of all life in the ocean. This is because the smallest of creatures consume them. These creatures are themselves the food source for other slightly larger creatures and all the way up the food chain. Located at the bottom of the food chain, these plants are critical to all life in the oceans. When we lose the basis of the food chain in the oceans, we lose the basis of all life on Earth. Aquatic fish, birds and mammals, and even land animals are dependent on a healthy oceanic ecosystem. These plants also work to produce a considerable portion of the oxygen for the atmosphere.

It has already been proven that the ever-increasing temperature of the Earth is likely to cause a rise in the seawater level. The catastrophe this would cause to our society is practically unspeakable. In fact, the global-average sea level has risen between 0.1 and 0.2 meters during the 1900s.[132] If it should continue to rise, freshwater rivers and lakes could be infiltrated by ocean water, terminating many sources of fresh water,[133] which is already in short supply. Major vegetation changes and the disappearance of forest species can surely be expected.[134] Plants and animals could become extinct.[135]

Warm episodes of the El Nino have been not only more frequent and persistent since the mid-1970s, but they have also been more intense.[136] Higher global temperatures will cause vast amounts of moisture to be added to the atmosphere.[137] This would cause snowmelts, cyclones, floods, landslides, droughts, and heat waves. Windstorms could run rampant throughout the planet.[138] Insects and diseases could increase in range and intensity. Rivers could turn into dried mud flats. Droughts could sweep across the Midwest; the entire eastern coastline could heat up; hurricanes of a size and proportion never before seen could pound coastal regions.

Spring is already arriving a week earlier in many parts of the world and forests are moving to higher latitudes.[139] The frequency and intensity of droughts have been increasing.[140] The World Meteorological Organization and United Nations Environmental Program all warn that we can soon expect heat waves much deadlier than any previously experienced. Climate models predict that if we keep increasing the emission of greenhouse gases into the atmosphere at the current rate, we will increase the global temperature by 18°F in just one hundred years.[141]

Insurance industries could go bankrupt from larger and more frequent disasters (weather-related insurance claims rose from $16 billion during the 1980s to $48 billion from 1990 to 1995).[142] But in 1998, violent weather cost the world over $89 billion, 48 percent higher than the previous record set in 1996 at $60 billion.[143]

In 1998, weather-related disasters killed about 32,000 people and displaced 300 million.[144] To put that number into perspective, that is close to the entire population of the United States. Upriver logging, wetland drainage, and climate change harks at the root of human suffering of unimaginable magnitude. The destabilizing effects could wreak havoc on our social and governmental institutions, as well as democracy and the very fabric of society itself.

Climate Flip-Flop[145]

Science has known for many decades now, and continues to find solid evidence, that the history of Earth is replete with wildly fluctuating climate changes. But only in the last 15 years or so have some of the reasons for these dramatic oscillations begun to be understood. To help lay people like myself appreciate how this mechanism functions requires some oversimplification of highly complex data from several specialized fields of knowledge including climatology and dynamic oceanography.

In the Atlantic Ocean there is a vast cycling of water that critically impacts the climate of the entire planet. To understand how this works requires a little patience as it can get a bit complicated due to the large numbers of cause and effect factors that influence this circulation, including the activities of human civilization. Nonetheless, we need to comprehend this predicament because the global warming phenomenon will eventually alter this vast cycling and recycling which has the capability of provoking a new ice age in the matter of just a few years.

The Gulf Stream, which is a warm flow of water from the Caribbean tropics northward up the American east coast, can be likened to a river of colossal proportions. This warm ocean current feeds into the North Atlantic Current which feeds the warm water up into Ireland and north up to Norway. This current is responsible for keeping Europe, and indeed the entire northern hemisphere, from freezing solid.

Once this warm tropical water reaches the northern parts of the ocean it begins to cool. The wind evaporates the surface water which raises its salinity, along with its weight. (Salt water is heavier than fresh water.) The combination of cooling and higher salinity makes it drop in great blobs of water. In the Greenland Sea and Labrador Sea this colder, heavier, saltier water mass sinks to the bottom (the process is called a downwelling) and then heads back south, looping back to the equator underneath the Gulf Stream. This loop is called a conveyor and the entire process is called thermohaline circulation.

This enormous "river" would fail to head north if the water up north did not sink, which would prevent the influx of more water, as well as the loop of cold salty water back south underneath. As has happened many times in planetary history, when this stream fails to head north, the Earth plunges abruptly into an ice age, at times for many thousands of years.

So the question becomes: What would prevent the water up north from sinking? We have learned that water will sink if it becomes heavy through increased salinity. Therefore, the surface water will not sink if:

1. fresh water is added, or

2. the surface water does not evaporate.

Here are four ways this could happen:

1. Insufficient wind blowing across the northern seas to evaporate the water;

2. Excessive rain adding too much fresh water to the mix;

3. Melting glaciers and/or flash floods of fresh water from ice dams breaking in fiords;

4. If fresh water is added to the surface water, the sea's surface could freeze (fresh water freezes at a higher temperature) which would prevent the wind from evaporating the surface.

These four events are made more likely to occur by higher global temperatures and any one would likely bring about another flip in the global climate, which is overdue anyway. As noted in the last subsection, glaciers are already melting. Fresh water lakes are spilling into the Artic region. It is actually incredulous that human industry is cultivating the very conditions needed to actuate an ice-age response from our biosphere. A climate flip-flop would be disastrous for a global population base of nine billion starving people, which would immediately descend into total chaos, not to mention pollution from freshly abandoned industrial and nuclear facilites.

We earlier discussed how warm climates can produce their own feedback loops which produce even higher temperatures. Likewise, abrupt cooling due to thermohaline circulatory failure could also produce feedback loops. An abrupt cooling would cause higher wind velocities and drier soil conditions which would result in an increase of sand being blown into the ocean. This sudden influx of nutrients would cause a rapid increase in ocean algae in turn creating a dramatic increase in the removal of carbon dioxide from the atmosphere. While that may sound like a good thing now, in this case it would be catastrophic because it would quickly add to the cooling rate of the planet. Also, heavy snowfalls would reflect heat back into space, causing temperatures to drop even faster.

Time for a Little Encouragement

While the magnitude of all this may appear to be completely hopeless, there are possible solutions and they will all appear to you as they each present themselves with absolute clarity. The simplicity and obviousness of each lucid point makes me, and perhaps you, wonder why these situations are not already being addressed. So be assured that answers to these dilemmas will all be addressed in Sections III, IV, and V. I certainly would not want you to become so despondent at this point as to give up hope! But all of this will come in good time. As you may see, I'm starting off with the bad news first. So let us continue to explore the effect our culture is having on our world.

Vanishing Wildlife and Dwindling Diversity

Perhaps the most obvious indicator of the perils of Domination can be found with the rapid disappearance of wildlife now evident on every continent and every ocean. Science has identified over 1.75 million species of life, although they admit that the true number is likely over 14 million. Periods of normal extinction rates over the history of our planet have been estimated at one every four years[146] to three per year. But now we are finding that the rate is currently up to 120,000 times larger than that as 30,000 species, at a minimum, are becoming extinct every year heralding the greatest extinction of plant and animal life since the age of the dinosaurs came to an end over 65 million years ago.[147]

Richard Leakey described in his book *The Sixth Extinction* a situation in Ecuador where ". . . ninety species of plants became extinct in a virtual instant, when the forested ridge on which they grew was cleared for agricultural land. The ridge, in the western Andean foothills of Ecuador, is called Centinela, and among ecologists the name has become synonymous with catastrophic extinction at human hand. By chance, two ecologists . . . visited the ridge in 1978 and carried out the first botanical survey in its cloud forest. Among the riot of biodiversity that is nurtured by this habitat . . . were ninety previously unknown species, including herbaceous plants, orchids, and epiphytes, that lived nowhere else. Centinela was an ecological island, which, being isolated, had developed a unique flora. Within eight years the ridge had been transformed into farmland, and its endemic species were no more."[148] This is not just an isolated incident where a local travesty occurred. This has happened thousands of times the world over.

In 1998, I read one of the most disturbing newspaper articles I have ever read. As the front page headlines were examining presidential sexual indiscretions, this small article was buried in the back pages:

Study: Earth's Resources Declining

GLAND, Switzerland—The planet lost almost one-third of its natural resources and animals between 1970 and 1995, the World Wide Fund for Nature said in a study released Thursday.

Freshwater environments such as rivers and lakes were the worst hit, with species they contain declining at a rate of 6 percent annually between 1990 and 1995, according to the 36-page Living Planet Report.

"These figures are a stark indication of the deteriorating health of natural ecosystems," said Jonathan Loh, one of the report's authors.

The populations of more than half of 227 species of freshwater birds, fish and mammals that were tracked had shrunk. Less than 10 percent had increased in number.

The report estimated that human pressure on natural resources is growing at a rate of around 5 percent per year. The figure is based on emissions of carbon dioxide and consumption of grain, ocean fish, wood, fresh water, and cement.[149]

Since life is so complex, most of what we are losing is unknown. But here are a few statistics that may surprise you:

- Over two thousand species of birds representing one-sixth of the world's species have disappeared from the Pacific.[150]
- Twenty-five percent of the entire world's mammalian species are on their way to extinction.[151]
- Whales, seals, dolphins, and porpoises are all in decline, sometimes dying simultaneously in herds of tens of thousands.[152]
- Over three thousand native species of plants have been doomed to extinction in Ecuador from 90 percent of its Pacific lowland and foothill forests plowed into plantations of oil palms and other exotic crops along with 54 percent of all its remaining forests.[153]
- In Alaska, fur seals, sea lions, and other related animals have had their populations decimated by warming ocean temperatures over the past two decades.[154]
- There has been an 80 percent decline in zooplankton, vital to the food chain, off the California coast since 1951.[155]
- One-third of all fish species are threatened with extinction.[156]
- After the Friant Dam was completed on the San Joaquin River in California in 1944, it took just five years for the salmon counts to plunge from 60,000 to zero.[157]
- Twenty-seven species of freshwater fish in North America have been driven to extinction.[158]
- Buffalo, elk, deer, pronghorn antelope, bighorn sheep, and moose have been reduced to only 1 to 3 percent of what their population was before the European invasion of the Americas.[159]
- One in eight of the world's known vascular plant species is threatened with extinction.[160]
- In the U.S., nearly one-third of all plant species are threatened with extinction.[161]

- At least 5,400 animals and 4,000 plant species worldwide are known to be facing extinction.[162]

- The majority of biologists believe that within the next thirty years, one-fifth to one-half of ALL species will become extinct.[163]

- Over one hundred Pacific salmon stocks have become extinct.[164]

Why is this happening? What are we doing that is precipitating such dismal statistics? We are:

- altering wildlife habitats;

- hunting them down in many ways, and very efficiently as well;

- polluting their homes;

- transforming and fragmenting ecosystems (forests, wetlands, rivers);

- utilizing vast tracts of land for intensive agriculture;

- converting rainforests into mono-cropped plantations;

- damming rivers;

- creating lagoons filled to the brim with pig and cow excrement as well as chemicals and toxic industrial wastes;

- creating cities and suburbs connected with land intensive and polluting transportation systems; and

- altering the weather.

Faltering Ecosystems

For every action, there will be an equal reaction. Environmental health is certainly no exception. Alterations in climate and weather patterns, ozone depletion, air and water quality, and human intervention is causing increased rates of tree disease and death. As large areas of forests die, not only will the animal life that once thrived there be annihilated, but also the forest itself can further injure and pollute the air by producing carbon dioxide through the wholesale decomposition of large amounts of organic matter.[165]

A consortium of Japanese research institutes ran computer simulations of what consequences global warming would have on the forests of the world, which forecasted some dismal predictions. They concluded that if the average level of global warming should climb by 3.5°C, 43 percent of the current global forests would be destroyed.[166] Usually climate change such as this occurs over a period of ten thousand years, and even then it would be considered rapid. What we are witnessing is the fastest catastrophic alteration in the Earth's ecosystem scientists have ever measured.[167]

Prove it!

Of course, the rate of change is debatable and some of the numbers and measurements are not known with great accuracy. Industry reps and their political puppets, along with rest of the corporate civilization apologists, take advantage of some of these gray areas and squabble about scientific precision. They fund their own studies that invariably show conflicting results, or pick on a few minor statistics that are irrelevant to the fantastic enormity and scale of the cataclysm humanity undeniably faces today.

This type of bickering can be likened to a fire in your home. What if you delay putting out a fire because you don't know how long it will take before it actually devours your house, or you can't quite accurately measure the flames, or can you measure the units of heat at the core first before you decide to take action, or exactly how much smoke is this fire producing? Of course, this type of hesitation would be insane, but it is just what we are doing now with regard to our environment.

Destroying the Web of Life

These horrors can be the only result of a system that alters the very basis of life. You can't destroy the habitat of a certain species and expect it, or the other species of life that may depend on it, to survive. From the change of sea plankton by the alteration of the ozone layer to the extinction of a small minnow in a stream destroyed by logging, these minor changes affect all life that depend on these smaller forms for sustenance. As the changes occur over and over, the web of life itself is altered and we may endanger our own species.

A perfect example of this is found with pollinating vertebra. At least 60 percent of all kinds of pollinating creatures have one or more species at risk of extinction.[168] But plant species are also at risk, along with the other forms of life that are symbiotically related in the dance of life.

Some might say, "Who cares? Something else will just grow there instead. What difference does it make?" The difference lies in the rate of extinction, and right now those statistics we examined earlier show us in a freefall. If life changes, it might be true that something else would exist in its place. But our culture fails to heed the fact that Homo sapiens are also part of life. We are a symbiotic part of nature, whether we are willing to admit that or not. It is quite probable that our species would be one of the species that would be replaced should a massive snowball of extinctions begin rolling. This is why we need to concentrate our full attention to bringing the impact this human population is exerting against our environment to a stop.

Of course, the reasons not to destroy entire ecosystems are lost on people and aggressive corporate interests who only see how they can make a profit on the exploitation of these critical systems. But even if an ecosystem or a species of animal or plant should not have the right to exist on its own merits and self-interests (from the viewpoint of human Domination), there is still immense value these components of our biosphere bestow on our home. Here is the "short list":

- oxygen production
- water purification
- solar energy
- waste treatment
- wildlife habitat
- flood prevention
- protection against soil erosion
- food production
- protection from sunlight
- regulation of atmosphere
- weather stabilization
- climate control

Effects of Domination 73

If you take an aquarium and add a chemical to it, all life within that closed system will likewise need to change and adapt. In the same way, once the actual biosphere we exist in totally has been modified, the contents of it will also mutate. In that same line of thought, the human species is transforming the web of life and the symbiotic relationship between not only individual species, but also entire ecosystems and all the species within them.

When viewed from space it becomes quite clear how fragile and thin the layer of life truly is. We are part of a delicious soup of life, and the master chef has been doing a fine job. Nature has been successfully stirring and mixing the right ingredients into this sphere of life for quite some time. Domination culture seems to think they can cook too, so let's examine some of the ingredients we are adding to the pot.

Toxins and Chemicals

The vast majority of the potentially neurotoxic compounds found in our environment have not been tested. The U.S. Environmental Protection Agency has identified 75,857 chemical compounds for use, with an additional 1,500 added yearly.[169] And we don't know what effect various combinations of these compounds might have on our bodies and the planetary ecosystem.

Polychlorinated biphenyls (PCBs) are all-pervading in our ecosystems.[170] Smaller organisms pick up chemicals from the soils and the waters. As they are consumed by higher forms of life, which are in turn consumed by even higher forms of life, the concentration of these deadly chemicals increase at each level. The higher on the food chain we eat, the greater concentration of this toxin we ingest.[171]

Heavy industry lobbying efforts combined with an aggressive propaganda campaign have influenced governments to unnecessarily fluoridate municipal water supplies (in the form of **sodium** fluoride as opposed to naturally occurring **calcium** fluoride[172]) based on the erroneous notion that the internal consumption of fluoride through forced mass medication could possibly prevent tooth decay. Sodium fluoride is a toxic by-product of iron, steel, aluminum, copper, lead, and zinc production, as well as many other industries such as glass, plastics, and gasoline. It causes damage to the livers, kidneys, and immune systems of both humans and animals. Children's teeth are prevented from forming strong enamel by its presence.[173] Sodium fluoride inhibits the valuable enzymes and vitamins that we need to maintain health.[174] It is not a human nutrient by any stretch of the imagination. Furthermore, the addition of this toxin to rivers through municipal discharges in the Northwest U.S. and British Columbia is now determined to be one of the main culprits in the high rate of trout and salmon loss in the Snake-Columbia River system.[175]

Chlorine or chloride is also added to our tap water to sterilize various impurities. Unfortunately, drinking these chemicals also serves to sterilize our intestinal tracts, which inhibits the natural fauna (friendly bacterial strains such as lactobacillus acidophilus and bifidobacteria, among many others), leading to the proliferation of Candida albicans.[176] This can cause infections in the upper respiratory tract, colds, cystitis, prostatitis, vaginitis, endometriosis, fungal nail, skin infection, athlete's foot, allergies, impotence, and depression.[177]

Dioxin, which is a by-product of the chemical processes that use chlorine, does irreparable damage to human and animal life, such as causing cancer, severely lowered immunity, and a disruption of cellular function. And now we can find this toxic substance in water, fish, produce, and meat.[178]

Methyl bromide is a pesticide that contributes up to 15 percent of the global warming gasses. Not only is it at least fifty times more destructive to the ozone layer than CFCs (by weight), but it causes neurological damage to the central nervous system and reproductive harm to all of the animal kingdom, including people.[179]

It takes almost two thousand years for rivers and lakes to clean out chloroform, another common industrial toxin. Yet this chemical is continually dumped into water supplies through treatment plants and spills. It is extremely toxic to animals and plants as well as humans.[180]

Pesticides are applied in work areas without the knowledge of the people who must occupy those spaces. Other chemicals recently introduced to our environment find their way into our bodies in several different ways. Toys, fabrics, perfumes, hairsprays, food, water, some cookware, furniture, carpets, cleaning chemicals, insulation, engine exhaust, and agricultural runoff are laden with chemical compounds that are not found in nature. Organochlorines, perchloroethylene, dichlorodiphenyltrichloroethane, methoxychlor, heptachlor epoxide,[181] along with thousands of other equally exotic sounding substances are now omnipresent in our environment. This blatant application of chemical compounds never before seen in nature makes us all unwilling guinea pigs in some mad experiment with natural life processes.

We all have accumulated synthetic chemicals in our fat cells, which naturally trap unknown molecules.[182] Many of these chemicals have the uncanny ability to mimic or block our body's hormones,[183] disrupting cellular activity.[184] Many of these chemicals are linked to specific degenerative diseases as well as deficits, such as learning and brain function.[185] Various compounds are biologically active. Some act like estrogen. Many jeopardize fertility. In fact, the average male sperm count has dropped 50 percent in the last fifty years.[186]

Oddly enough, we have increased the production of the most dangerous of them—the chemicals that have been linked to reproductive disorders. The average daily use of these dangerous substances between 1992 and 1995 was seventy-eight tons. The average in 1996 was one hundred tons.[187] Even though we have no idea of what long-range effects these laboratory-concocted chemicals may incur on life, one hundred tons of them are introduced into every sacred space of the living Earth every day.

Basic physiological processes have occurred naturally throughout eternity without artificial chemical interference.[188] In the last fifty years, we have spread these man-made chemical concoctions throughout the planet without the slightest idea what effect they may have on biological systems. We can only guess what the future holds.

Our bodies did not evolve with these substances. That is an undeniable fact. It has been theorized that because of this, we have no natural defenses to break these unnatural substances down and neutralize them. But we must ingest them because they are omnipresent in our "aquarium." No one can know what the long-term effect could be for future generations, and so we are gambling with the future of all life.

Genetic Pollution

One of the greatest problems lurking on the horizon of the future of all life on Earth is the corporate use of genetic technology to alter the genes of plants and animals. Genetic engineers are now splicing genes together from different organisms in unique sequences that are not found in nature. The purpose is to create properties in these organisms that make them resistant to various pests, lengthen the shelf life, improve the flavor, increase the yield, or alter them so that they produce toxins as their own pesticide.

All that may sound desirable, but there is a very dark cloud hanging over this new technology. First of all, scientists are now transferring genes between animals, plants, viruses, and bacteria. This will allow for allergenic properties from one form of life to be transferred into a completely different form of life. You might eat a chicken with peanut genes. You might eat a tomato that carries genetic material spliced from a fish. If you develop an allergic reaction, there will be no way to determine the cause.

But even more frightening than that is the fact that the genetic traits of G-E (Genetically Engineered) crops can be passed on to other plants by hybridization. This occurs as a result of wind dispersion or irrigation run-off. Once these genes become free into the ecosystem, they can mutate and reproduce in untold forms and combinations. Cross-hybridization could create super weeds and bacteria. Of course, as natural species are altered, the balance between species will be altered too, in ways that cannot be anticipated.

Unintentional cross-hybridization is genetic pollution. It is the spreading of altered DNA, the blueprint of all life, into the wild and natural world. It is quite probable that the genetic structure of all future life will be modified unnaturally. Once DNA is insinuated into a new organism, the change is irreversible. At that point there is no turning back.

We are creating the foundation of future generations, but this future could be a horrific existence with genetic pollution and inherited deformities. Only our descendants will be familiar with what the consequences of our genetic and chemical experimentation will be.

Corporate Agriculture

After the World Wars, new agricultural techniques were developed to increase production. Mono-cultured crops were developed and planted in the fields all over the globe. Petrochemical fertilizers and pesticides were introduced to further increase yields. Agricultural production became mechanized and automated, greatly increasing crop yields. Vast surpluses helped boost the population explosion to new unimaginable rates.

Family farms were very connected to the natural cycles of nature, rotating the crops and taking special care to recycle the nutrients in the soils. These farms would many times fail and would be acquired by larger farms. Gradually even larger corporations bought out these larger farms. Many times the land was purchased by companies that not only would market the food but would ship it, package it and provide the seeds, the chemicals, the fertilizer, etc. The decisions of how these lands would be treated came to be made by executives who had no direct contact with the land, but whose primary concern was the profit motive.

Food corporations began to proliferate which relied on monoculture, keeping the same crop growing in the same field, year after year. Unfortunately, this practice encourages pests. If a pest has an affinity for a certain crop and if that crop reappears yearly in the same field, the pests will multiply their numbers. Nature rarely if ever has produced an entire species that is a monoculture because they are easy prey for diseases and pests that have the capability to mutate and diversify. Once in, an invading species could easily take out an entire mono-cultured species faster than it could diversify to protect itself.

Crop diversity is the magic key to farming without chemicals. Gary Paul Nabhan in *Cultures of Habitat*, wrote, " . . . diversity minimizes the risk of total crop failure, since different crops respond in different ways to droughts, early freezes, insects, and diseases. In heterogeneous fields, some susceptible crops simply escape being seen, smelled, or otherwise located by pests. Faced with a mixture of susceptible and resistant crops maturing at different times, an insect pest is impeded in its evolutionary ability to overcome the resistant gene found in just one of the crop varieties present. Fields that are structurally complex also abound with pest predators, since hedges, vines, upright annual, and succulents provide nesting and perching sites for birds and other agents of biological control."[189]

Chemical fertilizers became popular in an attempt to replace what was now missing from the soils. Fungicides were introduced to prop up the genetically weak ones. Pesticides were used to kill the pests that were the result of the poor farming practices of monoculture. In this way the human food supply became devoid of the range of nutrients we require to maintain health. The nutrients in our food ultimately emanate from the soils of our food source. The lack of these critical components of whole nutrition leads to malnutrition and degeneration. We are learning that the quality of our sustenance is more likely to be the basis of health than the quantity. Nutritionally deficient foods create nutritionally deficient bodies prone to degenerative diseases.

Dead Soils, Dead Food

Humus is the organic matter found in all untouched soils, either living or dead, and is the food the plants take in through their roots. Plants are able to extract the minerals from this organic matter. Humus also enables soils to absorb and retain moisture, also critical for any plant's survival.

Live organisms find their homes in the humus, and can range from the microscopic to the visible, thus making soil a living ecosystem unto itself. Bacteria, fungi, actinomycetes, algae, protozoa, mites, various invertebrate, slime molds, mycorrhiza, earthworms, and nematodes all thrive in a health soil ecosystem.

All of this life is involved in critical processes that the plants depend upon such as decomposition, the processing of vegetable and mineral matter, the production of enzymes, and the recycling and detoxification of various organic substances that may be natural or synthetic.

Unfortunately, agribusiness is not interested in the dynamics of life beneath the surface as long as those aspects can be artificially reproduced to grow the plants as big as possible for harvest. The damaging effects of clearing, tillage, chemical fertilizers, and pesticides have had the effect of an exterminator. We have already lost over 75 percent of America's topsoil. In undisturbed conditions, natural processes will add one inch of topsoil every six hundred years.[190] Americans have eroded over fifteen of the twenty-one inches the European settlers found.[191] Precious topsoil has been hopelessly eroded as the life is snuffed out within them and 85 percent of this is directly due to the raising of cattle.[192] Our farmlands are becoming devoid of life as we add various chemicals and artificial fertilizers. Soils are so barren of life they turn to dirt. Their only purpose is to prop up our crops as we force-feed them enough artificial nutrients to make them look like plants.[193]

Normally, plants that humans consume would be able to thrive with no need for special protection from pests. But now that they are hybrids (genetically identical), they encourage the expansion of pests that thrive on those plants. The same fields see the same plants year after year and this encourages pests because they learn where their food continually reappears. They evolve to adapt to the poisons they find there, causing the chemically dependant farmer to introduce new and more potent poisons. Heavily financed laboratories create new compounds to keep unnatural chemical pace with natural biological evolution, which will always be the stronger of the two sides in that war.

The Department of Agriculture has determined that insects damaged about 7 percent of the crops in the U.S. before World War II, and that since those days that percentage has doubled to 14 percent. But the use of insecticide has increased tenfold in the same period.[194]

Corporate leaders think they can beat Nature at her own game. That would be laughable if it wasn't so tragic. Modern agriculturists apply insecticides, herbicides, and pesticides. "Cide," which means "to kill" is the root in all these words. The soils are poisoned and voided of the microorganisms that are needed to supply plants with nutrients. Our health is then compromised, not to mention our very survival.

In the real world, any plant will contain forty, fifty, or even sixty naturally occurring minerals as an integral part of its structure. Over-farming and poisoning the soils causes demineralization. To counteract this, agriculturists apply harsh petroleum-based chemical fertilizers, which usually consist of only three or four elements. That may well be all the minerals it takes to make a plant look like a plant—however, that plant will not be strong and able to survive without the artificial care a farmer will provide to make sure the crop reaches the market. Furthermore, the trace minerals are missing from the plants now. This further reduces structural integrity and lowers immune response because the raw materials the organism requires in order to perform critical functions are no longer available.

The genetic qualities of these plants are not bred for nutrition, but for size and other factors that make them easier to transport after harvest, e.g. tougher skins for tomatoes and drought-resistant corn bred for predictability, uniformity, and mass production. These genetically identical mineral-poor plants succumb easily to fungus, which naturally decompose weak vegetation that is unable to defend itself against infection. This causes the chemically dependent agriculturist to apply yet another "cide"—fungicides.[195]

Weak food will create weak bodies. The food is weak because it is missing many of the critical trace elements. The building blocks of our bodies are not available to us in these unnatural foods. Without the nutrients in the soils for the plants to ingest, the plants therefore lack the nutrients for humans to ingest. All plants now have much lower levels of minerals, amino acids, and vitamins than just a few hundred years ago. Our body's immune system needs nutritional support, and the plants we eat just do not contain the full range of nutrients. With a weakened immune system, we become susceptible to a host of ailments and degenerative conditions.

Poisons and Pathogens in Our Food Supply

Many poisons are introduced into our food supply with the aforementioned chemically dependent farming practices. Our vegetables and fruits come to us loaded with pesticides. Up to 1.5 billion pounds of these chemicals (some of which were originally developed during the world wars for the purpose of killing people) are applied annually to farmlands in the U.S.[196] The feed, which is given to the animals we eat, has less restrictions and so possess an even greater amount of chemicals. These chemicals wind up lacing the fatty tissues of those creatures.[197] Many of these chemical compounds can wreak havoc in the human body and cause a variety of degenerating ailments.

Why isn't the United States Department of Agriculture more concerned about dangerous agricultural practices? Is the public lobbying for growth hormones, toxic chemicals, and dead soils? No, but guess who is—international chemical and food processing corporations that have the financial resources to influence the policies and regulations of public agencies in the world's governments.

The animals of today are quite different than the animals that lived hundreds of years ago. Today they are genetically designed to be meaty, and are sad caricatures of their wild ancestors. While this genetic sameness makes raising these animals more profitable, it also means that they will be susceptible to the same pathogens. Microbes that successfully live in one creature will explode through an entire population, if those creatures are genetically identical.[198]

Flocks of chickens are confined in overly cramped factories. Their feed is contaminated with the wastes of other birds. Infections can spread rapidly. Rats and mice can carry the infections as well and can infect the next batch of chickens. Chickens that are enroute to slaughter are so terrified that they lose control of their bowels and spread their feces to other chickens.[199] The process of deboning and plucking chickens is totally automated and performed by machines. The machines rip, puncture, and otherwise violate the corpses of these creatures, spreading the contaminants locked inside their bodies from one carcass to the next. The process of ripping open the carcasses spreads the feces from the intestines around and imbeds it within the muscle tissues of the birds.[200] Also, the dead and diseased chickens that don't make the dinner table are recycled into feed along with their feathers, which have been marred with feces.[201] This is how diseases carry over from one generation to the next.

After being imprisoned, tortured, and slaughtered, the legless, headless, featherless, and gutless chicken corpses rest in a chilled bath for an hour. Why? One might imagine that it is to clean them. The reality is that this process adds 8 percent water weight to the flesh. It has been shown that any birds not contaminated before this point, will now be contaminated before being wrapped and shipped to the local grocery store.[202]

Up to 80 percent of chickens are contaminated with Salmonella,[203] 83 percent with Campylobacter,[204] and 99 percent of them have E. Coli bacteria.[205] Since muscles are known to be sterile it would follow that any contamination must come from feces.[206] Most people would be shocked to learn that there are more intestinal (coliform) bacteria in the meat preparation surface of their kitchen than on the rim of their toilets.[207] The final chicken product you see in the meat case of your grocery store might as well have been washed in the toilet. Chickens are the most contaminated products brought into modern day kitchens.

As contaminated chickens sicken up to eighty million people every year,[208] the general public is admonished by the meat industry and our government to cook our meat all the way through. This is how they pass the buck on to you. They provide you with a dirty diseased corpse and then say, "Hey, what's the matter with you? If you don't cook that meat all the way through, it will be your fault if you get sick."[209]

Beef cattle, pigs, and other tortured creatures we confine for our gustatory pleasures are just as sickly as the poor chickens. Most cattle are handled very inhumanely during the end of their years. They are also fed dangerous drugs, antibiotics and hormones—all of which you ingest when you eat the meat of these animals.[210] Just as an example, 93 percent of all pigs destined for the dinner table are fed a steady diet of antibiotics.[211]

As a matter of fact, most farm animals are fed antibiotics on a regular basis. While the initial reason was to ward off various pathogens, they also discovered that certain antibiotics enhanced the animals' growth.[212]

Animals are fed mega-doses of antibiotics to stave off disease and foster growth. As strains of pathogenic bacteria survive to reproduce, the meat we purchase can infect us with these strains that are becoming increasingly resilient to antibiotics. For example, Campylobacter is a very dangerous pathogenic bacterium found in many chickens. It also happens to be resistant to Bactracin, an antibiotic used in chickens. As Bactracin destroys microbes, the Campylobacter has just had its competition eliminated. Given the right conditions, it can reproduce and intensify.[213] As the antibiotic free-for-all continues unabated, super-strains of bacteria could soon evolve which would quite likely be resistant to ALL drugs and treatments.

Seafood has its own toxic brew to offer. Fish from the ocean are tainted with heavy metals, PCBs, pesticides, and a host of other chemical toxins.[214] Undersea nuclear tests have also contaminated the oceanic ecosystem and food chain with high levels of americium, plutonium, tritium, iodine 131, and krypton 85. Shellfish and crustaceans are loaded with a wide range of toxic pollutants such as methyl-mercury, PCBs, dioxins, and pesticides dumped by coastal urban centers.[215] These non-degradable compounds accumulate in the food chain and end up in the fatty tissues of the people who eat them causing a wide range of degenerative disease. Half of all fish, in a test conducted by the Consumers Union, was contaminated with bacteria from the feces of humans or animals.[216]

Farm animals eat the most appallingly unnatural diet anyone could possibly imagine. Animals bred for human consumption eat entrails, fetuses, condemned carcasses, feathers, bones, connective tissues, organs, blood, manure, litter, dead birds, and eggshells.[217]

Now the food industry is trying to push for the use of irradiation, another way to cover up their ineptitude. Irradiation will allow them to continue unsafe but cheap methods of processing meat so they don't have to be concerned about fecal contamination. Problem is, though, that irradiation does not kill 100 percent of the coliform bacteria. The fecal matter, along with some of the microbes, remains in the tissues. This environment is now ripe for a new microbe to infest the tissues that are free of competition, or will allow those that remain to multiply under the right conditions.

Hamburger can be a mixture of up to one hundred different cattle from two, three, or even four countries.[218] Grinding up meat is especially dangerous because that process spreads all the bacteria from fecal contamination throughout all the ground up bodily tissues.[219] The cows that make it to your dinner plate could easily have been infected with such conditions as pneumonia, tapeworm, peritonitis, and high fevers.[220] Cow heads oozing with regurgitated food (known in the slaughtering biz as "puke heads"[221]) have been found to get past hurried inspectors. These heads are destined to become hamburger, hot dogs, and luncheon meats.[222] Cows with high fevers,[223] leukemia,[224] peritonitis (a bloody, mucus-like fluid that collects in the cavity of the carcass[225]), pneumonia, arthritis,[226] and other nasty diseases all routinely make their way into the human food supply.

Toxic Ignorance

The public is not taught how to handle foodborne pathogens. We touch raw fecal impacted meat and open the refrigerator without thinking to sterilize our hands. We may forget to sterilize a utensil, and someone else will use it. This type of slip-up happens all the time in kitchens at home as well as in restaurants.[227]

The younger generation will usually be more ignorant of safe meat handling practices than older people. The younger workers, especially in the fast food industry, are the lowest paid workers. Many of them are poor and they come to work sick because they need the money. They haven't been properly taught how to deal with microbes. They don't really know how to safely prepare food. They don't know how to clean up and how to avoid contaminating other foods. The scary part is, if you eat fast food, they are probably going to be cooking for you soon.[228]

Cruelty to "Food"

We imagine the life of a dairy cow to be rather idyllic and relaxing, as if they're pampered and live the good life, at least for a cow. But it couldn't be any further from the truth. A dairy cow is kept pregnant all the time.[229] Sadly, her calves are taken away from her immediately after they are born. Dairy cows are milked against their will two to three times a day by machines.[230] Their udders become swollen and sensitive and frequently ooze pus from the sores into the milk. They are confined in cages where they cannot even turn around.[231] Even though their life span would normally be about twenty years, they lose their usefulness after about four, at which time they are ground up to make hamburger.

Cows are shipped by truck or rail to the feedlot where they are fattened up for slaughter. The journey, which can last as long as three or four days, is a veritable nightmare. They get trampled upon, and many end up with broken bones. They are not given water or food, nor are they given even the slightest comfort. Their handlers frequently treat them brutally as they are poked with sticks and electric prods. They spend the last four months of their lives chained at the neck while being forced to eat pellets made of other cows and animals. Some of the pelletized animals they eat are euthanized cats and dogs with the chemicals used to kill them still present in the bodily tissues. The feed they're consuming is loaded with drugs, antibiotics, and all sorts of chemicals.[232] Even chicken litter is in their feed as a cheap source of protein.

After that miserable experience they're back in the trucks or trains where they are carted off to the slaughterhouse. These smelly factories are the epitome of brutality, cruelty, hell, and corporate indifference to suffering, lacking any semblance to common decency whatsoever. As the cows fight for their lives, they can readily observe the cow before them downed by a stun gun, falling to their knees before they are chained by the rear hoof and hoisted in the air upside down. They kick and scream to no avail as their bones are crushing under their massive weight. And while this giant hideously terrorized and tortured mass of a being is seething with fury and writhing in excruciating agonizing pain, someone walks up and slits its throat while it is still alive so the heart will pump the blood out.

The life of a chicken is just as hellish. Crammed together in tiny cages in windowless factories, they live utterly miserable lives. Their little legs buckle under their abnormally large genetically engineered bodies. They peck at each other and live their entire lives in anger. Sometimes they lose it and pile on top of each other, smothering some of their fellow chickens.

Turkeys, as with all domesticated animals, have turned into gross caricatures of the majestic creatures they once were. When their time comes, they are hung upside down (while still living and conscious) and dunked in a bath of brine that is electrified before the conveyor belt moves them by a rotary blade that slits their throats (although sometimes the job isn't fully accomplished).[233]

The other white meat, however, comes from the most cruelly treated animal in the world—pigs. Pigs are raised in huge industrial complexes with populations numbering over one hundred thousand. The air in these sadistic torture chambers is thick with ammonia from their excrement. The deranged hopeless creatures are confined in narrow steel stalls over slatted floors, which are built over large pits into which their urine and feces fall. Their cages are stacked upon each other so that the excrement from the pig above continually falls on the creature below. Unfortunately for pigs, they have a highly evolved sense of smell. They are forced to smell this stench day and night. Pigs are very intelligent beings and have playful personalities and yet they literally go crazy under these conditions. Their final fate is the same as a cow's, murdered in the most brutal fashion beyond our imaginations.

Torture Chambers for Animals

Universities and government agencies perform cruel experiments on animals that are totally useless because the results are never conclusive.[234] This is because the physiology of animals is completely different than humans. There is no point in describing the gruesome details of animals that are clamped down while being subjected to painful ordeals. The point I would emphasize is that any animal experiments have no place in science. They're costly, provide incorrect conclusions, and are evil, in the most ghastly sense of the word. There are cheaper alternatives that provide more accurate results.

Degenerative Disease

There is no debate that the human diet has shifted in a radically new direction in the past few years. We're eating less vegetables, fruits, grains, seeds, and nuts while loading up on more meat, eggs, and dairy. Our food is highly processed which alters its composition, making vital nutrients to sustain health less and less available. Potato chips, pretzels, hot dogs, sodas, alcohol, and other snack-type foods were not on the menu of our ancestors.[235] White flour, polished rice, breakfast cereals, and the like contain only empty calories with the vital nutrients such as vitamins and essential fatty acids processed right out through the removal of the bran and the germ.[236] We're eating food that is canned, artificially sweetened, laden with sodium, defibered, preserved with chemicals, artificially colored with toxic substances, homogenized, and bleached.[237] Then we expect that these altered foods (by definition, drugs) will sustain us. How can they when the nutrients have been processed out?

Animal consumption, chemicals, and other toxic pollutants are creating new and exotic maladies. Multiple Sclerosis, diabetes, osteoporosis, allergies, cancer, heart disease, dementia, and arthritis are actually not diseases; they are a breaking down of the body. No longer capable of generating new life, the body simply falls apart at the seams on a cellular level at first, eventually manifesting itself in these grosser outward symptoms as the tissues themselves fail and breakdown.

It is an indisputable fact that humans live longer and healthier when they consume a high-fiber, low-fat diet that does *not* contain any animal products, including eggs and milk.[238] We have no nutritional need for any animal products whatsoever.

Homo sapiens are primarily herbivores. This is a physiological fact about the entire hominid line. Our body's digestive system is created for processing and extracting nutrients from a plant-based source. Studies have shown that the diet of early man contained very little animal meat.[239] The biochemical structure and nutritional requirements of the hominids was formed twenty to forty million years ago.[240] We descended from creatures that have the same basic digestive tract as we do. Their diet was 80 percent vegetables or more, so why is our diet only about 5 percent as such?[241] There are two good reasons for this colossal gustatory faux pas:

1. Animal fat is addicting. We evolved with this addiction because as wild humans we needed to be able to survive under the harshest of circumstances. When found without a food source, it was an evolutionary advantage to crave meat, which was quite difficult to obtain. Sugary foods were likewise quite rare, and we evolved with the addiction to encourage the ancients to consume it to ingest its source of fuel, even though in great quantities over long periods of time, these foods will be fatal.

2. Domination culture imagines the human species is the supreme highest of all life forms, representing the pinnacle of the evolutionary processes in the entire universe. As we imagine ourselves at the top rung of the evolutionary ladder, so we think that we must also be at the top of the food chain.

There are many mammalian species that have developed simple digestive tracts that are short and straight in order to quickly break down and eliminate animal tissues and intestinal contents. We're talking about animals such as dogs, cats, hyenas, wolves, and tigers. They all have hinged jaws and sharp, pointed interlocking teeth designed to rip open the bodies of their prey. Carnivores sweat with their tongues and have a very acidic saliva, lacking the enzyme ptyalin for predigesting grains. They have twenty times the hydrochloric acid in their stomachs, which break down animal proteins, than do herbivores. Carnivores all possess claws. The salivary glands in their mouth are very small.

Herbivores, on the other hand, have long intertwining intestinal tracts, replete with pockets and corners. These characteristics belong to such animals as cows, apes, horses, giraffes, monkeys, and chimpanzees—as well as us Homo sapiens.[242] None of these creatures possess claws. They all have very little hydrochloric acid in their stomachs, which is critical for breaking down animal proteins. Their saliva is very alkaline and contains ptyalin for grain pre-digestion. Herbivores perspire through millions of pores on their skin and have flat back molar teeth for grinding their food. Lastly, they are able to grind their food on these specialized teeth because they have an advanced jaw design that allows for side-to-side movement, something that no carnivore can do.

Protein is the word used for complex organic compounds formed by amino acids linked together in chains of peptides. Protein is what all life is made of. But in order for protein to be of any use to our bodies, we need to be able to first break those proteins down into the individual amino acids so our bodies can reassemble them according to our human specifications. Unfortunately, for those of us who have developed an addiction to flesh, the breaking down of animal proteins releases acidic compounds that enter our bloodstreams.[243] In an attempt to rectify this critical situation, our bodies turn to the store of calcium in our skeletal tissues to correct the pH level. This causes several degenerative conditions such as kidney stones (formed by excess calcium in the bloodstream) and osteoporosis (brittle and calcium-depleted bones).[244]

Cow's milk is created for cows—BABY cows. Adult cows don't drink it, so why should adult humans? Furthermore, dairy products wreak havoc on the human intestinal tract as the mucous coats and mats down the sensitive villi, which are responsible for assimilating nutrients. Dairy products also curd up around other food particles, preventing their digestion as well. The leading cause of allergies happens to be cow's milk. Cheese, ice cream, yogurt, butter, cream cheese, and milk deserve no place whatsoever in the human diet.

Furthermore, the ratio of phosphorus and calcium in cow's milk makes it difficult to digest properly. The optimum ratios are those where the ratio of calcium to phosphorus (Ca/Ph) is 2/1 or higher. Human milk has a Ca/Ph ratio of 2.35/1. But in bovine milk this critical factor drops to 1/27. In other words, milk from a cow has way too much phosphorus for humans—but not for cows of course, because is it created for cows. In the human's gastrointestinal tract, excess phosphorus combines with calcium and actually prevents the absorption of calcium. Therefore, humans actually absorb less calcium from the higher-calcium bovine milk than from the lower-calcium human milk.[245] But there is another reason why milk is a bad source of calcium: Bovine milk is loaded with bovine proteins. And just as is the case with meat, milk proteins also leach calcium out of our skeletal tissues.

Should children be drinking cow's milk to help build strong bodies? Absolutely not. All dairy products are pathogenic substances to be avoided at all costs. Children who are taken off milk from a cow usually become free from colic, colitis, earaches, colds, and congestion.[246] Sixty percent of the dairy cows in the U.S. are infected with leukemia. Their milk is a toxic brew of chemicals they ingested from their poisoned, rotten, spoiled, pesticide-insecticide-fungicide-herbicide-laden, and bacterially-infected food.[247]

There are a wide range of conditions and diseases caused by animal product consumption. Depending on an individual's personal physical strengths and weaknesses, the effects of animal consumption will vary. These facts are simply not debatable. Please share them with your family doctor. Let's examine a few of them:

- Meat and dairy load up our bloodstreams with cholesterol and fat. This causes **arterial sclerosis** (the lining of fat and cholesterol along the circumference of our veins, arteries, and capillaries). This condition leads to many degenerative conditions.

- Arterial sclerosis decreases the diameter of the blood veins causing an increase in pressure, just like squeezing a water hose, causing **High Blood Pressure**.[248]

- Increased levels of fat and cholesterol in the bloodstream decrease the amount of oxygen that can be delivered to our cells. Our cells need oxygen for the energy it provides in the creation of adenosine triphosphate (ATP). ATP is the molecule that forms the basis of every physiological reaction. More fat and cholesterol = less oxygen = less energy = **lethargy**.

- This lack of oxygen, along with crystalline deposits of uric acid (from urine) and the aforementioned excess calcium, causes **arthritis**.[249]

- Excess fat in the bloodstream alters human hormonal production. Excess estrogen in women stimulates abnormal breast tissue growth and cell division. This can lead to tumor growth and **breast cancer**,[250] as well as **ovarian cancer** and **uterine cancer**.[251] In men, excess fat increases blood levels of testosterone, which stimulates prostate tissue growth leading to **prostate cancer**.[252]

- Fat blocks our body's natural insulin from functioning, causing **diabetes**.[253]

- Eat fat, get fat. **Obesity** is a direct result of the overconsumption of all animal products.[254]

- Eggs and dairy have been identified as the primary culprits in triggering **asthma** attacks.[255]

- Putrefying flesh often becomes lodged in the human intestinal tract. This is because fat is solid at body temperature. Here it becomes carcinogenic (**colon cancer**) and punches holes in and creates grotesque outpouchings (**diverticulosis**)[256] that can rupture.

- Bovine Spongiform Encephalopathy (Mad Cow Disease and most likely Downer Cow Syndrome) is caused by feeding animals to animals. These degenerating conditions have been linked conclusively to **Creutzfeldt-Jakob Disease** (**CJD**) in humans.[257] Other viruses and pathogens are lurking in the meat we buy causing an untold number of cases of unnecessary degenerative and viral diseases.

- Excess cholesterol causes **gallstones** and **colon cancer**.[258]

- Excess dietary animal proteins rapidly deplete calcium from our skeletal tissues causing **osteoporosis**.[259]

- High animal protein intake greatly increases the risk for **non-Hodgkin lymphoma**, a type of **cancer** of the blood cells.[260]

- Bone calcium loss produces excess calcium in the kidneys causing **kidney stones** and **kidney cancer**.[261]

Even the **flu** and other dangerous **viruses** often originate in animals bred for human consumption. These viruses are carried harmlessly by birds or pigs. They may not have a deleterious effect on these creatures, but can be deadly in humans. Typically viruses, such as the flu, can change or swap DNA creating a new genetic code that, slightly altered, can be fatal for humans. The unnatural concentration of animals in farms can only increase the odds. Eating their bodies will also increase these odds. Many times, this is how viruses in a population can spread.

The Ultimate Cost: Poisoning Children

Corporations pay schools for the rights to push advertising propaganda of toxic products to students. Meat brands, irradiated products, genetically modified organisms, and caffeinated sodas laden with harsh simple sugars and chemical sweeteners that are unhealthy and proven harmful are hawked by ads, promote school teams and events, and are widely available in dispensing machines in the hallways and in schools lunches. Sodas are dangerous products, yet most people living in the U.S. drink more soda pop than the most essential liquid for sustaining life—water.[262]

Familiar cartoon characters push products that are unhealthy and dangerous for children. Foods laden with trans-fats, chemicals, hormones, and cholesterol are many times all that our children have available from which to choose. Now we find that obesity—a degenerative condition—is an epidemic among children today.

Medical Costs of Meat Consumption to Society

I have not heard what estimates may exist for the true cost of eating and drinking sugars and modified-for-profit foods. There are some estimates of what meat costs society in terms of medical expense. We've seen how the environmental costs are staggering. The other cost is the price we all pay for medical coverage. Estimates in annual health care costs in the United States alone, for conditions that are directly attributable to meat consumption, range up to nearly $61 billion.[263]

The Medical Model

Many diseases are merely symptoms of the body trying to detoxify. It is a painful process, but nevertheless, one that is necessary. The standard western medical response to most detoxifying situations is to cover up the symptoms so we feel better. Introducing even more toxic substances into the body is the method they have been taught. But as you plug up one exit for toxins, many patients tend to spring another leak. This creates a debt, over time, as the toxins continue to build up within the body and can eventually be manifested in degeneration of tissues and organs.

Degenerative diseases require a thorough holistic approach that examines all aspects of a person's life to determine what agent is causing the body to break down.

Unfortunately, medical doctors typically subscribe to the allopathic philosophy in which the practitioner attempts to merely relieve the symptoms of disease.

It is commonly known that medical professionals prescribe antibiotics (meaning "against life") quite often, even when they are not going to be effective. These powerful substances have the capability to sterilize our guts and kill various kinds of bacteria that might be populating our intestinal tracts. Unfortunately, antibiotics are very indiscriminate about which bacteria they destroy. We are supposed to have a total of over three pounds of bacteria in our guts. They help us break down our food, improve the absorption of fats and proteins, help with the retention of minerals, and prevent constipation, diarrhea, gas, and bloating. Many of them manufacture the B vitamin complex for our body as well as keeping the intestinal walls clean. They also work to keep out the bad bacteria we are trying to rid ourselves of. Harsh antibiotics sterilize all of these critical organisms leaving the intestinal tract open for re-colonization of newer and even more potent strains of dangerous bacteria.

Treating Symptoms, Physically and Globally

We live in a world where we deal with symptoms of distress, while ignoring the root cause. The allopathic physician is not nearly as concerned with how a disease condition occurred than with how to stop any indication of its existence. These physicians frequently use military terminology. The **war** on cancer, **killing** germs, **eradicating** viruses, **amputation**, **anti**biotics, and so on. These are typical Dominator responses.

And in other sectors of our society we have the "war" on drugs, or the "war" on poverty. We have the death penalty. It sounds as if we will always be at war, constantly doing battle with and always trying to exterminate our perceived competition. Got pests? Use pesticides. Got garbage? Incinerate or bury it. Need oil? Kill anyone who gets in our way. Need money? Levy more taxes. Have a problem with an addiction? Go to prison. Got pathogens in your food? Nuke it (irradiation).

Our society is always searching for new ways to turn our backs to why a problem exists and pretend it's not there by forcibly quashing its outward manifestation. We're throwing water on fires without doing anything to prevent them from catching in the first place. Unfortunately, as long as the wealthy are paid to cover up the symptoms, they will not be inclined to help solve the problems.

The Military Threat

The most obvious example of our society's symptomatic approach to deep-rooted problems is our propensity to resort to violence. Force, or the threat of it, is resorted to when human interests conflict with vested interests, corporate as well as political. This need, in turn, creates a whole new industry and opportunity for investors to reap huge profits through investing in corporations that provide goods and services to the global military establishments with unlimited budgets.

The U.S. accounts for 63 percent of worldwide arms. Most of these sales are to Middle East and other strife-torn regions where civilians are the major victims of war. These U.S. made weapons are used against our own troops by foreign militaries. This gives our government the excuse to fund research into more sophisticated weapons systems to overcome our very own exported technology. These sales drain capital from poor nations who should be spending their money on domestic programs. U.S. economic Support Fund Grants go to offset the costs of arms purchases. Many countries owe the U.S. billions in military loans, most of which are written off at the expense of U.S. taxpayers.

In the year 2000, the U.S. controlled half of the developing world's arms market with $12.6 billion in sales, according to an annual report published by the Congressional Research Service. Unfortunately, the U.S. and its corporations routinely sell these weapons to human rights abusers and dictatorships.[264] These are governments that deny basic human rights to its citizens and threaten neighboring nations. In the period of 1990-1999, the United States supplied many brutal regimes with arms through the Foreign Military Sales (FMS) program, or through industry contracted Direct Commercial Sales (DCS) programs. These sales included such countries as Algeria, Iraq, Lebanon, and Sri Lanka. The U.S. military, along with the CIA, have also trained the forces of many of these countries in U.S. war fighting tactics.[265] Our corporations right here in the good old U.S. have exported such useful military items as thumb cuffs, leg irons, and shackles to the tune of $5 million between the years of 1980 and 1993.[266]

World military expenditure in 2001 is estimated at $839 billion (in current dollars), accounting for 2.6 percent of the world Gross Domestic Product (GDP) and a world average of $137 per capita. This estimate is based on adopted defense budgets and is likely to be revised upwards when supplementary expenditures resulting from the September 11th attacks on the U.S. in 2001, and the ensuing war on terrorism, have been taken fully into account.[267]

Our corporate arms manufacturers are fanning the flames of war, while telling us about peace dividends. They earn big money selling bombs, guns, jets, tanks, landmines, and the list goes on and on. We're not helping to foster peace; we're encouraging murder and genocide. We are the merchants of death on this planet.

The military consumes vast portions of the world's available natural resources. Forty percent of the industrial plants in the United States exist to serve the military. War-related expenditures still account for most governmental expenditures.

Military Pollution

The U.S. military and other powers around the world have stored tens of thousands of tons of aging chemical weapons, many of which are leaking and could become highly unstable. Incineration has been shown to be a dangerous method of disposal, while other alternatives have been proven safe and/or cheaper. The military is nevertheless pushing for the construction of incineration plants.

Military operations by all nations, "friendly" and otherwise, pollute our planet with a pompous disregard for the places they exist to "protect." Air pollution from jets, ships, and vehicles; accumulation and dumping of hazardous wastes contaminating land and ground water; chemical accidents and weapons production, testing, and dumping, threaten many military hot spots here and throughout the world. Bases leak kerosene and gasoline into groundwater, and low flying jets disturb and destroy wildlife as the jet's shockwaves annihilate nests and habitats. In the early 1990s, it is estimated that the U.S. military generated more toxic wastes (up to 500,000 tons every year) than the top five U.S. chemical companies combined.[268]

The Nuclear Beast

This is the nuclear age. Military tests have destroyed life in many areas of the planet. But this is just the beginning, unfortunately, of what this nuclear nightmare is capable of.

Nuclear power facilities were touted to an unsuspecting public as a cheap and efficient source of power, infinitely available and as constant as it was reliable. Plus, it created no waste products or nasty plumes of soot. Just pure, ever-abundant electricity. The American home powered up, and consumer appliances started to fill the local department stores promising to make life easier for the housewife in performing her duties.

Nuclear waste, however, turns out to be an incredibly dangerous and horrific phenomenon that our society is not prepared to handle. It seems that there is no safe place to put these by-products. They're dangerous to move, last for hundreds of thousands of years, leak, and they're a growing menace to public safety.

Specialists and scientists have so far been unable to come up with a safe and permanent method of storing nuclear wastes.[269] High-level radioactive waste is piling up in irradiated fuel pools. The industry wants the public to foot the bill for a permanent storage site in a remote area. This would require over 15,000 shipments over the next thirty years by trains and trucks. This is simply not feasible because just one accident could potentially kill thousands and injure millions of people, costing us up to $6.2 billion to clean up.[270]

Chernobyl released nine billion curies into the atmosphere killing over eight thousand people and increased the background radiation of the planet by a factor of 180. Strontium-90 and cesium-137 were released causing an increase in leukemia and thyroid cancer. Its future legacy is expected to include inherited genetic defects such as deformities and other developmental abnormalities, some of which won't manifest for two or even three generations. Genetic mutation of our children and their descendants is the price we may all be paying for cheap abundant power from "Reddy Kilowatt."

Of course, we rarely hear about this because the media is owned by the nuclear industry itself. General Electric and Westinghouse own their own television networks. Don't expect to hear much unbiased news coming from them.

The propaganda from the industry tells us how cheap nuclear power is. Just throw those elements in there and presto, electricity comes out for practically nothing. Nothing could be further from the truth. Since 1950 the nuclear industry has received over 100 trillion dollars in direct and indirect subsidies from the federal government.

Too Late to Change Our Ways?

Let's suppose our descendants decide to create a new world system based on subsistence farming, cooperation, communal living, and a reduced population base. Let's say they decide to value human and community relationships over national, political, or corporate ties and focus on their local needs. Or, what if the planet actually did experience a sudden flip-flop climate change? The human species was certainly able to survive through them before, but with our global population numbering in the billions, mass famine, starvation, civil strife, and war would surely be the result. In these situations and others we may not have thought of, who is going to tend to the waste sites? Who can safely shut down the reactors?

Or, what if we, as a species, decide we don't want nuclear power generators or the wastes they create anymore?

The problem is that the wastes, at least according to today's technology, will be toxically viable for many years. Some estimates can range up to 250,000 years. This mandates a military presence to prevent any hapless souls or other groups of unknown origin from accidentally or purposefully disturbing the spent fuel in its slow decay. Power plants, if suddenly abandoned for any host of unforeseen reasons, would go critical and life as we know it would be forever threatened. Our species along with the rest of creation would disappear.

And what if the governments, or even humans themselves, are not around in 200, 1000, or 10,000 years? What future pain are we creating for life that we couldn't possibly know at this time?

Forced Migration to Cities and Borders

As we have learned, corporations mine for minerals, drill for oil, and clear forests for lumber or to create pastures for grazing cattle. International corporations, to grow feed crops for animals, are buying up more and more farmlands that once grew crops for human consumption. Mono-cultured human food crops are also displacing locally based subsistence human food farmers as global governmental officials and corporations take their lands to grow cheap food for international markets. "Free" trade agreements make indigenous lands easy to grab by the rich and the powerful. In doing so, they annihilate native cultures while polluting and spoiling their birthright homelands.

Rural peasants are continually displaced as their subsistence style of agriculture is replaced with cattle ranches or huge corporate mono-crop farms for export. Nontraditional crops, both for food and feed, destined for the global market require large-scale agricultural techniques that are available only to the wealthier farmers. New seeds, equipment, and knowledge are required, along with tons of chemicals. As land becomes less available for traditional use, it results in higher prices all around for food and land. This prices the poor out of the market and their homes.

Livestock are devouring the vast majority of the crops grown throughout the world. In Mexico, millions of people are chronically undernourished, yet one third of the grain produced there is being fed to livestock.[271] In Brazil, soybeans for the far more profitable international feed market (food for farm animals) are rapidly displacing the peasant diet of black beans.[272] In Costa Rica, 80 percent of the arable land (land that is suitable for growing crops) has been converted to cattle pastures.[273] As David Morris wrote in *The Case Against the Global Economy,* "Brazilian per capita production of basic foodstuffs (rice, black beans, manioc, and potatoes) fell 13 percent from 1977 to 1984. Per capita output of exportable foodstuffs (soybeans, oranges, cotton, peanuts, and tobacco) jumped 15 percent. Today, although some 50 percent of Brazilians suffer malnutrition, one leading Brazilian agronomist still calls export promotion 'a matter of national survival.' In the global village, a nation survives by starving its people." These are but a few examples of the international corporate encroachment on rural families in developing nations, and the continued First World assault on and destruction of foreign communities and cultures.

Corporate agri-businesses around the world are continually clearing forests and forcing to extinction indigenous and peasant populations. This pattern is being repeated throughout the planet to raise cattle in support of the fatty diets of the wealthy in America, Europe, and other rich nations. Social unrest and political upheaval follow in the wake of the vast cattle complex as it rolls over the continually displaced poor. They can't fight it. They are not organized or supported in any way to protect their way of life. It is estimated that over half of the rural families in Central America, constituting over thirty-five million people, have been forced off their lands.[274]

But where do they go? They must lose the only way of life their families may have known for hundreds or thousands of years and move to the urban slums or shantytowns and attempt to eke out a living, seeking whatever employment they might find. This is probably the greatest mass human exodus of all time as our culture impoverishes the vast majority of humans, robbing them of their way of life and their sustenance.

NAFTA brings the trucks rolling and crisscrossing the continent from one market to the next delivering huge corporate profits while displaced Third World citizens work for slave wages. Some choose to leave their country where corporate and political alliances force them off their native lands and into slums. This has resulted in many mass migrations all over the world. In the U.S. we are most familiar with the "illegal" immigration of poverty-stricken Mexicans who must cross the borders to find work or create a new life.[275]

As these people are used for their cheap labor, they are also villainized by our culture for creating problems with our welfare system. Does anyone actually believe that these people (trapped in their own country after being forced off their lands due to cattle grazing and feed grain production for our hamburgers) would actually prefer abandoning their homes, families, and way of life for the squalid existence that is forced upon them in our country? As the powerful state-supported agricultural and financial corporate interests of the northern hemisphere take credit for the modernization of the southern half, they are destroying the traditional patterns of food production. In these societies it is the local economy and locally produced food that maintains social cohesion and cultural stability.

Free Trade Induces Dependency

The importation of cheaper foodstuffs from other regions or countries prices local farmers out of business. When this happens a dependency on trade for that food is created. Trade in today's modern world is based on oil, which is a finite and diminishing resource. Once a community becomes dependent, should it later lose the ability to purchase a commodity, it must rely on state or foreign aid to survive. This introduces further influence and control from foreign state and commercial interests. As they gain control of foreign economies, they can create markets for the export of their own products. As the once self-sufficient economies of the South become desperate to pay off loans, they destroy the lives of their citizens by turning agricultural lands over to transnational corporations, which convert them into export produce for cash. Oddly enough, they then must import food to feed these communities, which have now lost their very source of sustenance.

Foreign loans to corrupt officials can create debt that accrues interest, which allows and encourages vast extractions of local resources to pay off these loans, much to the further detriment of the powerless local indigenous cultures. Those who attempt to fight back will always lose as large shipments of armaments disguised as "aid" help foreign-backed dictators to violently suppress the will of their citizens.

Human Population Explosion, Food Shortages, and Famine

At the present time there are over six billion of us. We are the first generation to witness a doubling of our numbers. Our agricultural system is peaking in its ability to increase production to feed the population. Most of the human population haven't enough to eat and are dying from famine and drought.

Most of the population growth is found in urban settings, making an ever-burgeoning percentage of humanity dependent on agri-businesses for food. In 1995, forty-five percent of us (2.5 billion people) lived in cities, up 30 percent (734 million) from 1950.[276] The danger here is that urban populations consume and generate wastes, but produce practically nothing that would help to sustain life.

As our numbers increase, competition for water amongst our cities, industries, and irrigation does too. But as we divert an ever increasing amount of water to our cities, we are finding that less is available for the farms that provide the food to support those numbers, increasing reliance on imports and driving the price of food ever-higher, along with rapid depletion of groundwater tables.[277]

We are presently adding 90 million people to the planet every year. If the rate of population growth were to remain constant (it is still climbing), there will be 135 billion people on the planet in 200 years. But that is not possible. Rough estimates by researchers indicate that 11 billion is the maximum the planet can hold without triggering massive population implosions by way of war, disease, or global famine.[278] The planet does not have the resources to support our present numbers, much less this growth in population.

Stressed Out

Poverty, overcrowded populations, destitution, and loss of personal control creates mountains of stress for the modern human across the globe. As one might imagine, the stress in our reality is much different than that with which we evolved. Our ancestor's stressful experiences were seldom and short-lived when they happened. These experiences were usually due to immediate physical danger that required fast action. But in today's reality, the stresses we experience are low-level and constant. "What is the boss going to do?" "This traffic congestion will never clear." "The kids are driving me crazy." "How am I going to pay these bills?" Issues like these can extend for months and even years, slowly eating away at us.

Surely lack of time and money are two of the greatest modern stressors that our ancestors didn't even think about. We don't have the time in our culture to play with our children or relax for a couple of hours any time we feel like it.

The Time for Money Trap

Mostly we stress about money issues, and this can affect our relationships as well. Should we buy a new car, or clothes and a washing machine? A vacation, or a college education? House repairs, or medical help?

The need for cash translates into the need for a job, which translates into the sacrifice of time performing tasks to benefit someone else. The need for income is critical in our culture, which only allows those who have cash to eat and live safely separated from the elements. We can't gather food anymore, unless we want to dig in trashcans. No, we need a job. We need to get (give) hours. The more money we need, the more time we have to sacrifice.

It used to be that one income would suffice for a middle class family—eight hours a day, five days a week, two weeks off a year. But now we find that as the job market becomes vulnerable to unstable market forces, the breadwinner needs to work fifty, sixty, or even seventy hours a week. Some families in need of cash will find two working parents absent from the home and children being raised by day care facilities.

Family Alienation and Absentee Parents

All this leads to families losing their bonds that were at one time powerful and intimate. Parents, grandparents, children, siblings, cousins, uncles, aunts, and close friends are losing the glue that at one time made these relationships enduring.

Not only are our jobs and careers alienating us from one another, so is our predetermined private space. If you own a two-bedroom condo, there isn't going to be much room for a grandparent and/or an uncle down on his luck. We send family members whom we are unable to assist to nursing homes or homeless shelters. What else can we do? Families are cramped as it is.

Television is another great alienator, which we will be taking a closer look at in the next section. Night after night, times that could be spent with family members become times spent staring at the tube.

Many families nowadays only have one parent, and that parent is usually more stressed out than the families with two. " . . . Trends like unwed motherhood, rising divorce rates, smaller households, and the feminization of poverty are occurring worldwide."[279] The unnaturalness of the nuclear family confined to small rooms in a single unit usually results in dysfunctional relationships. The intense focus this forces on individuals brings out the worst in the human. A child's parents will often break up to keep the peace. Without other close bonds to support us, the modern relationship can easily go stale, and in many cases, rancid.

Legal Drugs for Socialization

Our ancestors used to gather together every day to socialize and play games. It used to be that hanging with everyone was a daily event. We miss it because we instinctively crave it.

At home it's just the same old routine, same old two or three people. We hook up our brains intravenously to network television for variety. The other answer of course is to go out and meet, dance, sing, and talk with other people in a fun-filled atmosphere, just like we did naturally when humans lived in tribal societies. Of course, many people now must go out to nightclubs if they want to socialize. And in nightclubs most people drink alcohol in order to socialize. In modern times, bars have replaced these ancient tribal customs that we evolved to participate in.

Drugs

The innate desire of all humans to alter one's consciousness is all-pervasive. The tribal rituals and rhythmic drumming and dancing created an altered state that helped the tribe coalesce into a unified entity. Gasping in awe of the magical world around them, people used to feel a spiritual connection to life. Losing that element in the modern world, whitewashed and sterile, people crave that out-of-body sensation that we evolved to experience.

Religious services offer caricatures to true natural spirituality, which serves to artificially recreate that lost element of the natural human psyche. Another method to help the mind escape the boring mundane repetitive nature of the workaday life is a wide variety of natural and synthetic drugs. Drugs can help lift the spirits of the depressed and calm the nerves of the irritated.

This is the reality we must come to expect until we learn new ways of helping people cope: People will get the drugs they crave. Alcohol, one of the most potent drugs available, was illegal during the prohibition, and the underground crime network provided it. As with the prohibition, so it is today. By criminalizing substances that people are willing to pay for, you invite the criminal element to participate in the distribution of these substances.

As a matter of fact, criminals want drugs to remain illegal. If drugs were legal, the criminals would have no market. The law enforcement community also wants drugs to remain illegal. If drugs were legalized, they'd have to lay off half of their employees. The corporations want drugs to remain illegal, otherwise they would have no market for the technology and weaponry that both sides of "the war on drugs" require. Over 80 percent of the increase in prison population between 1985 and 1996 were from the use and sale of drugs.[280] The builders of the prisons, the court system, lawyers, and corporate employers of inmates—they all like the status quo. It is no wonder that the privatized prison industry has been reporting record profits.

The loser is the taxpayer who funds this evil scheme to pilfer from the public coffers. The loser is the innocent citizen at home or out in the community who is robbed and assaulted for drug money. Addicts are driven to get the cash to pay for expensive drugs to reach their nirvana, which is something we are all genetically programmed to crave. It can even be a higher motivator than food. Forcing addicts to turn to the black market only gives criminals power and cash.

The illegal trade in narcotics has a captive market of about 190 million addicts and users worldwide, and is estimated to be worth more than 400 billion dollars a year. This makes it larger than the oil and gas trade, larger than the chemicals and pharmaceuticals business, twice as big as the motor vehicle industry, and second only to the world's arms trade, which is estimated at more than 800 billion dollars annually. According to R.E. Kendall, Secretary General of Interpol, "The estimated turnover of 400 billion dollars has the power to corrupt almost everyone."[281]

Drugs for Children

We unknowingly teach our children to use drugs. Children are force-fed commercial glamorization of alcohol, caffeine, and nicotine. They are taught to use pills to feel better.

And not only is our society pushing drug use on our mass consciousness, our corporations are hooking children on drugs. Caffeine is an additive found in large doses in most soft drinks—up to 50 milligrams per can. Artificial sweeteners and sugar can alter a child's mood and disposition. Artificial flavors and colorings, chemicals that alter brain chemistry, are also drugs.

Many children have a difficult time concentrating in school and on written homework. The causes of this malady can include excessive chemicals, lack of micronutrients, or heavy metals in their diets. Synthetic chemicals during pregnancy can disrupt the neurological development of the fetus. The medical answer to the problem is drug therapy. They call it "drug replacement therapy," but drugs like Ritalin are not replacing anything natural.

Our society values uniformity. All children are now called "students," and should behave as expected. If a student exhibits unique behavior that is disruptive in a classroom, that child may be given a diagnosis of Attention Deficit Disorder, or some other label, when the child is simply being a child. Indeed, the reason many children are drugged is to keep them under control in crowded classrooms.

Youth Gangs

Children who grow up in an atmosphere of dysfunction and absent parents may learn poor social skills. Social structure of a tribal nature, which all humans are genetically inclined to crave, is practically nonexistent. This is where street gangs fill in the missing gap.

Disaffected youth looking for guidance can find it readily available in the local peer group in their neighborhood. Easily swayed by the promise of being recognized and feeling important, these youth learn to work the streets together and make their own way in the world. They are emasculated and become leaders among their peers.

These people deserve to be leaders. All humans deserve this. It is our birthright. Our hierarchical culture denies us this by ripping families apart and creating a second-class citizenry and calculated unemployment to keep wages depressed.

Crime and Punishment

As a means of income, crime is all that appears to be available for some impoverished groups. If you need to provide for your family or just want to have money, crime is always a lucrative profession. Gangs, jails, and inner city groups offer the free training necessary to be successful. In the United States alone, we have over 5 million people in prison or on trial, probation or parole. This makes the U.S. the largest penal colony in the world.[282] The actual number of people behind bars is around 1.9 million.[283]

The prison industry is fast becoming privatized, with its own lobby, a mandate to grow like any other industry, political influence, and propaganda machine. Between 1990 and 1994, this sector of the corporate world grew by 34 percent,[284] and since the early 1970s, the increase of the jail population is 800 percent.[285] In 1998 there were 5.9 million Americans who were potential prison labor for corporations to tap for cheap labor.[286]

This corporate practice to enslave and employ has global implications as well. In his book *Unequal Protection*, Thom Hartmann wrote, "Under the new WTO and NAFTA rules, an importing country cannot consider the conditions under which a product was produced. So, some corporations have discovered that they can profit by using prison labor to manufacture export products or perform services for offshore clients." Prison labor can be found for next to nothing and so some nations work with the corporate sector "by passing laws that are easy to violate" such as "criminalizing health problems (drug use), or criminalizing 'anti-state' behaviors"[287] such as protesting, practicing religions, or engaging in political activities that are not sanctioned by the state.

The FBI has determined that street homicide kills 24,000 Americans every year. The Department of Labor has determined that 56,000 are murdered every year by the corporations they work for, either from occupational diseases or accidents due to poor safety standards. And while robbery and burglary may cost society about $4 billion in any given year, white-collar crime by our doctors, lawyers, accountants, and well-dressed businessmen costs $200 billion annually.[288] More times than not, these criminals get off easy.

And the Poor Get Poorer

The percentage of poverty-stricken families is increasing every year, and now nearly one-fifth of Americans live below the poverty level. One out of every five children in America are living in poverty. Even employed Americans are living in poverty because wages for the lowest economic tier are not enough to live on. And the gap is growing: the average wages of the lowest and the highest 20 percent in New York City are $10,700 and $152,350, respectively.[289] Average wages are shrinking as the wealthiest 5 percent continue to enjoy upwardly spiraling incomes. Most people want to work, but are not able to find jobs because:

1. U.S. jobs are pouring out of the country because of free trade agreements such as GATT/WTO and NAFTA, and

2. the system is structurally set through interest rates to keep a certain percentage of the working force unemployed. This helps to keep wages low.

Many people who cannot find a job need public assistance to feed their families, and then attract the blame from those who would label them as lazy. Amazingly, there are so many things in our communities that desperately need to be done; there is plenty of work to go around. The money system is simply inadequate to serve the needs of our communities by rewarding those willing to work with a job, and providing workers to fill the needs of the community.

In 1995, over thirty-nine million Americans were without health insurance, up from thirty-one million in 1985. There are now forty-five million uninsured people in the U.S., and twelve million of them are young adults who would like an education. At the close of the 1990s, 30 percent of young adults were uninsured, compared with 16 percent of children age 18 and younger and 16 percent of adults ages 30 to 64.[290]

Families with teenagers and young adults looking forward to getting an education must many times put it on hold if a family illness brings economic hardship. Those who are fortunate enough to get a higher education often find themselves going deep into debt taking out student loans to pay for it. And yet we find that one out of three graduates must take a job that does not even require a college degree.[291] In fact, the single largest employer in the U.S. is now Manpower.[292] Using temporary services, corporations who layoff their employees can rehire them or other suitable replacements without being required to pay any benefits and can let them out the door on a whim.

And yet, collectively on a global level, Americans are doing quite well, but that is unfortunately at the expense of the rest of humanity. The standard of living that Americans have come to expect and demand is possible only through the degradation of the environment and the quality of life in CIA-controlled Third World nations. The U.S., along with the rest of the industrialized nations, consumes 86 percent of the world's aluminum, 81 percent of the paper, and a whopping 75 percent of the energy. Incredible as it may seem, this is all consumed by only 21 percent of the human population.[293] Eighty-three percent of the global income goes to the richest fifth of the population. The poorest fifth receives but 1 percent.[294]

And while the developed nations take and take from undeveloped nations, what they give back is far less than ever before, especially since the termination of the Cold War when foreign aid was tied to political allegiance and corporate cooperation. Fifteen percent of U.S. taxpayer money went to foreign aid in the late 1940s, and by 2001 this had dropped to merely 0.1 percent.[295] While fully developed Western nations bask in the benefits of the natural resources of poorer nations, almost half of the people on this planet live on less than $2 a day, one billion of us do not have access to safe water,[296] malnutrition runs rampant, and tens of thousands of children die from disease and hunger. The U.N. has estimated that about $9 billion would help secure a safe water supply for the entire planet, which is roughly what is spent on cosmetics by Americans every year.[297]

Pointing Our Fingers at Each Other

Most Liberals blame the Conservatives, the Wealthy blame the Poverty stricken, the Educated blame the Uneducated, patriots blame the foreigners, every race has at least one other to blame, and it's all vice versa—it's a three-ring circus of fault-finding and false accusations.

While everyone is jockeying for the microphone to utter the next sound bite to take the public's attention away from the facts,

- the forests are dying;
- poverty and famine haunt billions of souls;
- the atmosphere is filling up with toxic gasses;
- the oceans are becoming sterilized;
- inner city crime and violence is increasing;
- the rivers and streams are drying up and filling with silt;
- the weather patterns are changing;
- the human experience is trivialized; and
- species are disappearing.

All the public gets is simplistic linear talk for complex multidimensional and interconnected facets of a phenomenon that spans the entire globe.

Our wasteful endeavors even extend out into space itself with space exploration programs. Can a lander on Mars do anything to justify the costs, economically as well as environmentally, or the risks of failure, which is a common occurrence? Is this *really* going to increase the quality of life on Earth, or in actuality, is it just an effective method to fatten the wallets of those who own the technological corporations that contract their services to the governments involved?

Looking Ahead

It's been quite a journey, but the journey is far from over. I hope this section has served to make abundantly clear where Domination is leading us. To me it seems that the human species is utterly incompetent in managing the planetary environment, and even our own culture. What can we do to stop this downward spiral? Does it have too much momentum? Has the bell tolled for our species as it has for so many other species and cultures? Can those institutions that are eroding the quality of life all throughout the planet be controlled? Can enough of our cultural siblings become aware soon enough to allow us the time to find and enact the solutions?

As frustrating as it may be, knowing what the problems are is the most important step, because it is the first. The second step is learning what is creating these problems. In the next section we will study the mechanisms and operating systems that turn the wheels of civilization.

III

Aspects of Domination

> *"In modern Western history, power and influence have traditionally been shared and/or balanced between the four 'estates'—the government, business, academia, and the media. Today . . . money is controlling all four . . . "*
>
> —Bernard Lietaer

Pods, Troops, Prides, Packs, Schools, Colonies, Flocks, and Tribes

Wild killer whales (orcas) live in social organizations we call pods, which normally number from five to thirty individuals. Pods are matriarchal societies with one dominant mature female. The males usually leave their pod and become solitary. From time to time they will temporarily associate with pods that contain potential mates. Sometimes pods will join together to form herds of up to one hundred individuals.

Baboons on the other hand live in a society that has four levels. The smallest unit at the bottom level is a group that consists of one male that has between one and up to ten females in a harem system. Perhaps there may also be a few non-breeding males as well. Female offspring will eventually leave the group to join other harems, while males attempt to attract or kidnap available mates for their own harem. Related males join their groups together at various times of the day in larger groups called clans and they forage for food together. Clans will also form even larger groups called bands that will also forage for food. Finally, several bands come together at night to sleep in an organization known as a troop.

Then there is the lion, which lives in groups we call prides that consist of two to eighteen females with their offspring, and from one to seven males. Once again, prides are matriarchal societies. They do most of the hunting and make all of the decisions such as where to sleep, eat, and find water. While females and the female relations will spend their lives in the pride, males are temporary. They get ejected from their birth pride between two and three years and form coalitions with other males and become nomadic. These nomadic male groups look for prides to overthrow and become the new resident males of an established pride. This can happen to a pride as often as every three years.

The gray wolf lives in social organizations called packs. Packs normally consist of five to eight members. Generally, these consist of a dominant male and dominant female with their offspring. The pack is organized in a caste system. First are the alpha male and female, next are the non-breeding adults and last are the social outcasts that live on the periphery of the pack. Juveniles do not enter this system until they are adults (about two years of age).

Ants live in colonies where there is a very strict caste system in place. A sterile caste has the smallest individuals taking care of the nursery, and the larger ones acting as the colony's heavy workers and protectors. It is through the type of nutrition they receive in the larvae stage that determines their body size and shape and what role they will play and behavior they will exhibit. Bees and termites also have different classes of a division of labor.

The talapoin monkey lives in groups that consist of three main subgroups. One is all adult males, another is all adult females with their dependent offspring, and the last is the juveniles of both sexes. There is little interaction between the subgroups, and even within the subgroups themselves. Individuals do not mix with others and only associate closely with one to three others. When females are ready to breed, they will join a group of males higher in the forest canopy.

Certain species of fish form leaderless schools that maintain perfect synchronicity swimming in parallel, turning and fleeing at exactly the same time. They almost appear to be a single organism. Flocks of certain bird species also exhibit this same ability to act as one.

I cite these examples of natural organizations of wild animals to make only one point, but a very important one. Just as wild baboons naturally form in troops and wild ants naturally form in colonies, wild humans naturally form in bands and tribes.

Individual tribes may have totally different and varied cultures. Most tribal societies typically have around five hundred members, which are broken down into smaller units called bands consisting of twenty-five members.[298] This is certainly not the rule, but a general tendency. (Other studies suggest slightly different numbers, but this really isn't all that relevant to the point I'm making.) The bands of a tribe share the same language and dialect. A tribe remains in a general geographic area that they become quite familiar with. But this familiarity is more than just knowing their way "around town." We use maps and pinpoint our homes, the stores, the family, the church, the schools, and the "job." We know which routes to take to get us there. But indigenous tribal members know their ancestral homeland in such a deep, comprehensive, complex, and holistic manner, we would not be able to fully understand it without devoting our lives to it.

Tribal societies are immersed in the web of life that lives and thrives around them, and which influences their culture. Their connection enables them to truly understand the high degree to which their lives depend on this web. Their respect for the web is reflected and acknowledged by their culture. Information about the plants, the seasons, the animals, the land, and how to live is handed down by oral traditions.

Most are egalitarian, not hierarchal, having no one particular leader, depending on the circumstances or the group task at hand. In some tribes, one or more of the mutually accepted members with certain specific skills rise to lead as the situation calls for it. They know who is best at finding food or building shelters or communicating with other tribes or settling internal disputes. They know who is best at going to war or healing the sick or injured. While we view all of these tasks as jobs that require positions of "authority" that need to be officially delegated, their culture does it for them.

Each culture has its very own method of breeding customs. Some are matriarchal, others patriarchal—and still others may have other systems of social organization we might not understand. In some tribes the men might have several wives, while in others women may have more than one mate. Some have complex rules that govern their selection of mates between bands within their tribe or maybe from neighboring tribes. But three things they all have in common are:

1. a spiritual connection to the land upon which they live and their culture evolved,

2. a spiritual connection with each other (ensuring each member of the equal right to live and belong), and

3. a spiritual connection to the tribal sense of purpose.

These are the basic values that are evident in most every natural tribe or indigenous culture. They feel like a part of the land they live on, not that the land is their property. They naturally care for each other as a unit, as they are each and all one with the tribe. Everyone automatically cares for the young and the old. They don't feel that the tribe is an organization or an institution that owes them anything; everyone is the tribe. They feel that who they are as an individual is not as critical as is the tribe of which they are a part.

The land, people, and tribal unit are all one and the same. It is natural. It is the tribal way. It is so because the edicts of their culture have been handed down for thousands of years and are self-evident. It is so because this is how humans evolved to be.

Mankind is Kind, Man

I come from a family that is half-German and half-Norwegian by ancestry. When we were kids, my younger brother was learning about the German atrocities in World War II. I'll never forget when he made the realization that he himself was half German. He couldn't accept it. He was eight years old and could not believe that people who were capable of such evil actions could be even remotely related to him.

It is true that a very small percentage of our population are prone to commit evil deeds such as murder, rape, torture, and inflicting pain with conscious intent. Keep in mind that these people are members of our culture. However, I cannot accept any theory that espouses the belief that humankind is defective or flawed.

When I greet people face to face in everyday life and am able to make eye contact and truly communicate with who that individual is, it's hard for me to imagine that any human being could consciously perpetrate evils such as the holocaust or the wholesale destruction of a forest. How could any of us kill or intentionally inflict pain on any other creature, human or non-human, and view that action as necessary and acceptable? Indiscriminate murder is not a human trait and is completely antithetical to human nature.[299]

Why is there all this violence, greed, and suffering in the world today? Where does it come from? How can it be that human institutions have committed the atrocities we find in our history books and in current world affairs? How can any person with a conscience commit murder or torture with no remorse or misgivings?

It's Humane Nature to Nurture

If most of us view a baby animal, our first impulse is to stroke it. We humans delight in feeding, playing with, and caring for other creatures. It's inborn. This tendency is the result of a genetic stamp on our genes, honed through the ages. We like to care for other creatures and love them.

This urge to empathize is inbred. It's as much an instinct for humans to nurture as it is for birds to migrate, or babies to suckle. To be **humane** is to be **human**. We really are, in every sense, loving, compassionate, and caring creatures.

The supposed dark side of humanity is not something we are endowed with. We are, however, gifted with a huge complex cerebral cortex that allows our mind to be programmed by the culture in which we live and identify with. This flexibility to go beyond instinct with the power of our minds to willfully control and alter our thoughts and actions allowed for a fantastic ability for our kind to be able to adapt and survive in a wide variety of conditions and environs.

The current global culture we live in evolved and had to impart a philosophy that would allow our cultural ancestors to flourish—a new philosophy that was antithetical to our humane impulses, and yet it spread like a virus. It spread because domination and exploitation was this culture's primary features. A group like that, set next to a group that just wants to be alone, is bound to conquer the more peaceful neighbors.

And so it is now that the hunter/herder values that evolved over 10,000 years ago are taught to each one of us by our mother culture. We learn to distrust. We learn to exploit. We learn to compete until the losers and the winners are clearly defined. We learn to expect special treatment. We learn that we are the Chosen. We learn to regard our kind as having been given the whole of creation for our own exclusive control, profit, and exploitation.

The New Human Environment

I would suppose it would be very difficult for anyone nowadays to attempt to live as humans evolved, that is to say, out in the "wilderness." First of all, it no longer exists. But regardless, we view the natural world as the great "other" out there without the modern conveniences upon which we are addicted and have come to expect. But what price does this lifestyle engender upon the quality of our lives, or could it be that modern society is the superior way of life after all? Are we living in the pinnacle of human evolution?

Most people are rarely cognizant of the fact that today's humans are isolated from the environment that our species evolved in. We are sequestered from the very environment that we were designed to interact with. We did not evolve driving automobiles down crowded freeway corridors. We did not evolve eating hamburgers and French fries. We did not evolve sitting and watching four hours of trivial television every day.

Life in our reality is totally manufactured. Everything we see, touch, and interact with every moment of our lives is a product of human creation. The ground is paved. We live out most of our lives in cubicles we call apartments, condos, houses, churches, schools, stores, malls, offices, and vehicles.

Our senses interact in our homes and gathering places with conditioned air, electric lights, and sounds emanating from speakers, machinery, and traffic. Our minds focus on operating machines and appliances as we are constantly interacting with a myriad of gadgets, implements, and utensils. These things comprise our reality and truly shape how we think and what we know.

We travel around sitting in various types of metal boxes rolling at high speeds through a labyrinth of roads, streets, and freeways. Our reality here is concentrating on the speed and direction of the car while reading billboards advising us of things that are, in large part, bad for us or that we, at the least, have no need for.

As our species has become trapped and confined in these artificial environments, our relationship with the planet is warped. We look at an intersection of two streets, but cannot see the forest that once stood there just a few years earlier. We look at our buildings, living rooms, and corridors as if that's what always was. But where dead structures stand, just a few years prior, was once a living ecosystem.

We've lost touch with the natural cycle. We no longer observe a bud metamorphose into a fruit. We no longer notice a caterpillar transform itself into a butterfly. We no longer sit and are mindful of the sky and the subtle nuances of the weather and how it affects our environment. Oh, we can give it a quick look on occasion, but we don't have time to truly live and be connected with any of these things on a deeper, spiritual level.

Most of us grew up learning that apples are purchased at the store, and oh yes, they do grow on trees somewhere. We can't be truly concerned about how the apple came to be as long as we know we can just drive to the supermarket and fill our cart up with as many as we can afford. Our focus is on whether or not we can afford the apple. So we just go to work and earn money to pay the rent and buy groceries and then throw our garbage into the garbage can with little thought of where products come from or where they go when we discard them.

Work Environment

To get food and shelter most people in our society work in the business world. This is a world of desks, computers, and paperwork where the mundane repetitive tasks repeat themselves continually without pause. Offices and manufacturing facilities are brightly lit and generally without many windows. The whole idea here, of course, is to eliminate any outside sensory stimuli that could possibly divert one's attention away from the task at hand. In this sense, an office space could be considered a sensory deprivation chamber.

Our schools, which must prepare the adolescent human for a life functioning under these conditions, resemble offices in their detachment from the full range of human sensory capability. Here our children are forced to channel their thoughts in an unnatural way so they can emerge from these institutions and move into the business world with the required skills to serve.

As humans must perform in these environments devoid of spontaneous expression and interaction with life, the business of paperwork and strenuous mental exercise gains a seemingly larger importance. High grades in school and worker performance and profitability become the new standards by which we measure our individual worth and value as a member of our society.

The New Human Condition

Any being that is living in an unnatural state will exhibit unnatural traits that are not commonly found amongst those similar beings found in its natural habitat. A tiger in a cage, for example, will just pace back and forth all day long, totally different from a wild tiger. None of the creatures we hold in captivity for human viewing pleasure behave as they would if they were living in the wild. They develop odd habits and strange ways of relating to other creatures. They become caricatures of themselves.

In the rearranged reality we have created we have become dependent on whatever means we are provided with for survival. Our natural human responses to life situations are no longer valid. The fight or flight response, mostly useless in modern civilization, is one good example that helps to explain why there is so much stress in modern life. Our body's natural urges to react are continually suppressed because attacking your creditor or boss is not acceptable behavior.

Our expectations of what is possible, in terms of our very humanness, are vastly reduced by what our culture's expectations are and what our society demands of us in order to survive. As we attempt to fit this unnatural mold, we are not fulfilled in what our innate desires are dictating to our inner selves.

This conflict then expresses itself in a multitude of dysfunctional symptoms that our society labels as odd. The misfits amongst us are those who cannot adapt. They are unable to find a place in our society because there is none for them. They may have certain talents and abilities that might have had some use in another time, but here they are not marketable. They exist on the fringe with no idea how to merge with their surroundings. They live a meager life or turn to crime or other such unsavory methods of survival.

> *So, what is the big trade-off anyway? How could things be better than we have them right now? We have our jobs, our cars, our homes, our nightlife, our families, and our entertainment centers. How could living this natural life as wild humans possibly be any better? They didn't have the modern conveniences we do; they had to rough it all their life, forced to do battle with the elements and fight for survival in the unforgiving wilderness.*

Well, yes, if you took any modern day city folks and put them in the middle of a dense forest to fend for themselves, they'd probably be dead in a matter of weeks. But the same would hold true for an indigenous person thrown in the middle of our society as well. The same would hold true for any domesticated creature that never knew its roots. It wasn't raised there. Like us shopping at the grocery store, the tiger in the cage is thrown hunks of meat; a domestic dog looks to its dish.

But the survival issue is not really the point I'm getting at here; it is life experience. More specifically, it is the quality of our life experience. Humans evolved experiencing billions of multi-spectral sensual (on both the physical and spiritual planes) inputs that were interconnected on every level on a continuous basis. All the senses were constantly linked and utilized to interact with and merge human beings with the natural world. Humans evolved in an environment where they interacted with a myriad of other life forms. Life itself, in all its richness and complexity, enabled them to continually live fully through all six senses.

We, on the other hand, are severely cut off from that. What we touch, see, smell, hear, taste, and intuitively feel is limited to the grosser simplistic limitations of what elements are present in our preconceived reality. This, in turn, shapes our perception of life. The degree to which we are intimately familiar with various aspects of life determines and limits what we are capable of perceiving and understanding. The sacrifice is a deeper sense of belonging to the community of life, along with a connection to and a caring for the planet. We lose touch with those senses that are no longer useful, and in so doing, lose part of what we are capable of experiencing.

As our daily experiences become repetitive and mundane and we are chained to our routines, we become bored. In turn, this creates chronic exhaustion and we turn to compulsive and obsessive habits to stimulate our minds, or numb them. Food, drugs, money, sports, religion, violence, and career advancement become all pervasive in the empty spaces that were once filled with a constant communion with the web of life.

The Bushmen of Africa live under material conditions we would consider to be absolute poverty. Yet they appear to those who would research their way of life to be immensely content and happy—far happier, in fact, than the typical American immersed with the riches we have plundered from the rest of the planet.[300] Why does it seem that we, surrounded by our plentiful bounty of riches from all around the entire world, are sadly numbed and isolated within the four walls of our lavish homes, totally unaware of the extent to which we are deprived of the fullness and richness of our own sensory capabilities?

Human Beings, or Consumers/Employees?

We live so much more differently than our ancestors did, it would not seem to be the same planet. We only know our own personal experience, but rarely think much of the transactions we perform to stay alive. It wasn't always so that men had to leave the home to go follow someone else's orders eight hours a day to provide food and shelter for his family, although that's what we envision when we think of the typical family. But we did not evolve to live that way. It wasn't always so that people would go grocery shopping if they needed food, although that's all we know. Just a few generations ago, most people grew their own food. It wasn't always so that people were guided by a market economy's constant voice advising them about what they should desire, although that's all we know. We find out, more often than not, that what we find ourselves desiring are things we really don't need. Desire and Need are two different concepts, but our culture teaches us that they are the same.

Commerce used to be a personal thing that emphasized relationships more than maximizing profit margins. People would share the wealth that belonged to everyone. Trading and sharing between individuals, groups, families, tribes, and communities created human bonds. In the Domination culture, however, commerce has been reduced. Commerce is no longer a relationship or a bond between human beings; it is an impersonal marketing of products and services to consumers. I believe Freedom means the freedom to live one's own life in accordance with one's own personal values. But sometimes it seems that Freedom, in our consumerist, market-driven society, is just the "freedom to choose between brands."[301]

Commodifying Experience[302]

Our society/culture takes human creations and transforms them into products that can be duplicated and marketed. This is called "commodification." The purpose of this endeavor, of course, is to sell as many units as possible for a profit that can be squirreled away, and so the people must be encouraged to consume more.

How do they "encourage" us to consume? Advertising. Since corporations own and/or control (buying airtime) the media groups, they have the ear of an entire society and the global culture at large. This is the biggest cultural experiment ever performed on Earth. Human beings from every corner and every walk of life in every stage of life receive thousands of daily messages from a mere handful of corporations. And they only have one message: CONSUME. That's it! Nothing too evil sounding at all. Consuming stuff is what we consumers are taught to do. We are taught that happiness is derived from the quantity and quality of the stuff we have at our disposal. Even our cultural values revolve around consumption.

So, off we go to work to get money to buy stuff that we consume and use, and discard the extraneous stuff (batteries, packaging, etc.). We even toss out the very stuff itself, once its usefulness has been exhausted. And when we feel empty once again, we repeat the cycle.

To maximize profits, the corporate sector does everything it can to create demand for products by controlling the society and influencing the culture within which resides its consumer base. The very act of planning and mediating culture makes it shallow, almost non-existent by definition. In impersonal and anonymous transactions, the connection between individuals is lost. True culture is reduced because we end up interacting with our stuff and our work, and not each other.

True human culture thrives on continuous spontaneous interaction. It's the stuff that goes on between people and groups of people. But we're taught that culture is something that is presented to us by professional artists such as painters, musical groups, movie directors, television actors, and such who are financed by corporate entities that control the cultural message and content of the expression.

But, sadly, in this way it just becomes one more thing for us to consume. We may not realize how odd it is that we spend money to purchase culture and cultural experience. But by passively observing it and not interacting or influencing its evolution, cultural experience merely becomes a vicarious spectacle, something to observe and internalize.

We see that Domination culture has, is, and always will be mediated. It is not a spontaneously occurring phenomenon of creative human interactions, but rather an orchestrated presentation of human creations that are designed to control the actions of others.

Homogenization of Culture and the Promotion of the Consumer Lifestyle

When a company makes a product, it wants to sell as many of them as possible. They must convince the members of the society to buy their product. They must encourage those members to value the accumulation of products in general. They must promote the lifestyle that supports the principle that commodities bring pleasure.

Cultural values held by the society where any particular marketplace is found must be molded by the corporate community to sell as many products as they can. These "corporate values" are only effective if as many members of a society as possible hold these values as their own. Corporate values include the concept of the nuclear family living by itself in its own rented or owned house or apartment. Every family unit needs one of each machine used in the typical domestic dwelling. These families should buy the same foods, cars, diapers, beds, appliances, computers, and furniture. They should read the same newspapers, celebrate the same holidays, attend the same churches and schools, and drink the same beer.

Anyone who does not value these things is suspect. Diversity is viewed with fear or disgust by corporate values. Corporations look to create a sameness in everyone. This homogenization is critical to the creation of as large a marketplace as possible. The media, especially television, is one of the best tools around for this purpose. Most people think television is a reflection of our culture. They believe the process of watching television is a cultural experience, one that is merely made convenient through the advance of technology. Is this the people's culture? If it is broadcast globally, are we all now experiencing the same culture? Whose culture is this?

Modernists versus Traditionalists

Before we can answer that, we need to look at the cultures that are available for broadcast. For these purposes, I'm not referring to ethnic subcultures at all. For most of recent history there have been primarily only two subgroups of the Domination culture, worldwide. The first one, the Modernists, who have been in the driver's seat for the Domination culture since the Industrial Revolution, are mostly oblivious of the unnatural state in which they live. They accept the system as is and have complete faith in it. The other group, the Traditionalists (who have been on the scene for the past 130 years), is fearful of change, as is inevitable with Domination. But instead of working with the situation as is, they're constantly attempting to roll back the clock to a more familiar time, be it five or fifty years back. These two groups have been struggling to lead the world to their way of thinking for the past century. We will look closer at these two subcultures, plus an exciting new one in Section IV.

TV Land

While Traditionalists have good representation of their subculture in the media, the Modernist agenda rules TV Land, without a doubt. Slick advertising and programming continually reinforce the notion that all is as it should be. Featured products and lifestyles are coupled with talking heads delivering "news" and opinions, all of which mostly endorse the lifestyle Moderns experience. Since Moderns do not question authority because they trust it, they soak in their cultural marching orders four to five hours a day, spending vast amounts of their allotted existence in the Universe experiencing life through the lens of a television screen.

Unfortunately, television is *extremely* limited in the sensory data it is capable of transmitting. We look at the coarse picture and listen to the coarse sound, but all the other senses are basically dead in that experience. We can only experience what is dished out to us on those two sensory inputs. We are limited in the total sum of that experience, not only because of the crudeness of the sensory data, but also in what we are *not* experiencing, such as odors, feelings, atmosphere, tastes, touch, 360-degree vision, and depth of field.

We have little choice in what we are experiencing because we do not participate in how the information is collected. We can merely sit back and absorb the stream totally and fully, or cut ourselves off from it entirely.

People think that if they view a forest on television that they have experienced that forest. While they know that it is not a forest per se, on a deeper unconscious level they still believe they have experienced reality. This false sense creates a detachment from the natural world because the experience of it on television just wasn't that great. In fact, a rich, diverse ecosystem teeming with life can come across on television as rather boring.

The loss is not only found in the quality and quantity of the sensory information that is presented, but also the range of emotional feelings that the technology is capable of delivering. Subtlety in human emotion and expression is lost as the medium can only transmit the larger grosser aspects. Competitiveness will win over Caring and Cooperation. Violence and Fear over Peace and Love. Domination over Coexistence.

This is why sports work so well on television. It is constant rough and tough action. Police brutality, crime, car chases, soap operas, murder, conflict, bombs dropping, and sensationalistic news and drama all fit this basic simplistic mentality. But what about the subtle nuances that exist beyond the scope of a 30-minute presentation? I'm referring to such concepts as true caring for life and other beings, tranquility, tenderness, affection, and concern. How about the deep feelings of connection with life that one may feel on the shore of a pristine lake at sunrise?

It's all boring on television. Television must be loud and in your face every second to hold our attention. If it weren't, the very boring nature of the medium would cause our minds to wander away from it and defeat the purpose of the producers who need to sell their products. The danger here is that constant expression of gross emotions and feelings will be manifested in the viewer, especially children.

Brainwashing, Mind Control, and Alienation

As the images invade our conscious and unconscious minds, we lose connection with other humans even though we are in the same room. When a television is on it is next to impossible for humans to interact with each other because to do so requires them to break away from the stream, causing them to lose touch with the experience they are seeking. Television demands your attention. How many families function on bare minimal levels because the television is cutting them off from relating to each other?

Television alienates our children from their very humanness. Children spend many hours of their precious life watching "the tube" and interacting with video games. But if children are separated from other children and their parents by hours of television, video games and computers, when and where are they supposed to learn to be human?

The goal of television is to create consumers who value commodities. It is here that our children learn to be isolated and desirous of things that only money can buy, and that's just what the producers of children's programming are hoping for.

Physiological Impact of Television

But far beyond the direct implantation of a writer/producer's obsessive thoughts into the mind of a child is the critical effect of the medium of television and video games on a child's developing nervous systems. Children are exposed to this mind altering technology in ever growing numbers and percentage of life experience.

Since dawn immemorial children learn how to be a member of their culture by emulating that which is found in their mind's environment. It's how humans learn to be human: See, observe, and emulate. But as we have learned, television is completely limited in the range of human emotions it is capable of expressing. By watching and interacting with non-human television and video games devoid of life itself, children ingest the stream of images flowing into and altering the structure of their thought patterns and actual physical state of their brain. This is the great modern tragedy of the reconfiguration of the cerebral cortex's neuronal networks.

Furthermore, as they watch and experience these mediums, they intuitively feel the compulsion to physically react. But before their brains send a signal to their bodies, another signal from the conscious mind stops those commands from being manifested. They continually vibrate back and forth between action and repression; yes/no, go/stop, do it/don't. They're caught up in a trance-like state feeling this nervous vibration until release.

Once released from the state of mental suspension, they exhibit inappropriate behaviors as they attempt to return to their real surroundings. Socially unacceptable behaviors such as sudden bursts of energy and physically emulating actions their minds have just finished ingesting are what come naturally to them. If children spend vast amounts of their formative years immersed in and mesmerized by television, then the emotions and behavior patterns presented to them there will be the ones they will be learning, no matter how good their parents' parenting skills might otherwise be.

Mental Simplicity Breeds Control and Power over Public Opinion

The very nature of television disconnects the viewer from life and effectively channels it into its own simplified reality devoid of all the nuances that real life has to offer. This simplifying of reality has that effect on the mind of the person who merges with it. As they become passive and are drawn in, they tend to believe what they are witnessing is real. This is because humans evolved to believe what we see and hear. We think we are experiencing first-hand information, and so we tend to accept it as truth. This unusual arrangement allows the viewpoints and opinions of the producers and the sponsors to be directly implanted into the minds of the viewers.

This gives an incredible amount of power to a very select few. Though we experience television individually, as we must, what we fail to overlook is that millions of others are also experiencing the exact same thing individually. This creates a new phenomenon in human society: We are all walking around with identical memories; we're mental clones of each other. The power this gives our institutions, political and corporate, in the form of vast mass consciousness programming and control has no precedence.

Television has become one of the most powerful tools for political and corporate propaganda. Television viewing is recognized as a sort of hypnosis. The human mind gives itself over to the stream of consciousness that flows into our sensory inputs. Can you just imagine then the power of the ability to hypnotize millions of people in one singular market economy?

The top 100 advertisers pay for two-thirds of all network television[303] and only a handful of corporations decide what we view. They basically control our culture, molding it to fit their specifications.

Television thwarts democracy in that only those with the economic power and the social agenda of consumerism have access to the medium. Democracy does not exist in any society where the ideas that are transmitted to that society emanate from a select few.

The media regularly and willfully suppress information that would implicate the business practices of its advertisers and sponsors. The "news" we watch on television or read in mega-corporate publications is filtered to allow public access only to those facts that would not incriminate either the owners of the news outlets or the advertisers. Sponsors will also determine the content of other programming to ensure that it falls within their corporate guidelines and messages.

A Powerful Vehicle for Powerful People

The media is controlled primarily by privately owned corporations, which are financial vehicles for the personal profit of an elite group of people who make up a tiny percentage of the human population known as major stockholders. These individuals wield enormous power through the ownership of the corporate entities that are creating the fate of all life.

In Section I, we briefly touched upon the Boston Tea Party and how it was a protest against the corporate dominance of the British Corporation, East India Company. This led to the Revolutionary War and the birth of the United States. This nation was founded upon the freedom from corporate aristocracy, and yet the corporate control of government and society is without a doubt omnipresent throughout the U.S., as well as the rest of the world. It was considered by the founding fathers of the United States that the underlying philosophy of democracy is to protect the people from unrestrained corporate power. But now we find that corporations are the super rich and powerful global forces whose self-interests and lust for financial wealth are unrivaled in all human history as they control the flow of information.

Corporate Profit at any Cost

All manufactured goods require pieces of the natural world: animals, forests, minerals in the Earth, land, or oil deep below the surface. Corporations extract these pieces and process them into something else. If a resource should become unavailable or difficult to extract in one part of the planet, a corporation will simply move its operations elsewhere, even to a foreign country.

Corporate capital is utilized to manipulate foreign governments in many different ways to allow a corporation to move in and extract natural resources. Multinational corporations conspire with corrupt Third World political leaders the world over to rape the lands, pollute the waters, murder the opposition, and enslave or uproot the people who live on the land they want. This is done for mining, drilling, farming, or manufacturing of natural resources for consumer products, most of which is destined for the United States or other western nations.

This is a cost that is borne by humankind and all the inhabitants of the planet. The greater cost is borne by the future inhabitants of that locality that have *permanently* lost the resources (clean water, forests, and farmland) to a handful of corporate investors. A forest that may be home to a people is useless if a corporation destroys it.

Corporations inflict hierarchical form on today's culture, which puts some of us in an unnatural position of immense personal control over other people's lives. This is also found in the military, government, and religious organizations. Hierarchy is not an innate human trait. It creates stress, abuse, and aggressive behavior while discouraging cooperation and a caring mentality.

A corporation's stockholders may not be cognizant of the suffering their corporation may be perpetrating on society and Earth, even though they own a piece of it. As long as their investments are producing profits, most stockholders remain totally ignorant of how their corporations create wealth. Some may not even care.

Currency

Currency is the blood of the corporate machine. Most people don't even question the type of money they use in their daily transactions. We learn early in life that money can be traded for things and services, but few people really know what it is.

Modern paper money began as a note from a goldsmith that you owned a certain amount of gold in his safe. It was a promise to pay. Gradually governments gave banks the right to create money, which was backed by the gold, but this can be a bit of an oversimplification. (However, the Nixon administration took the U.S. currency off the gold standard in 1971.)

Banks are allowed to print more money than they have, as long as the government has access to whatever funds it needs. If a loan is created, the money is likewise created by the bank, as long as that bank has 90 percent of the value of the loan in reserve. The bank gives this new money to the consumer who passes it along to a party that is selling whatever it is that the money is secured to purchase, such as a home, car, or business. The seller in turn deposits this money into another bank, or even the same bank but different account. The bank where these funds are deposited is now likewise permitted to issue yet another loan of 90 percent of that amount, which in turn perpetuates the cycle, which in turn creates more fresh money. And on and on it goes. When any given loan is paid back, the money disappears back into the void from whence it came, but the bank now has the interest.[304] Lending institutions suck the wealth from the system in this way as they slowly accumulate it, transferring capital into the pockets of the shareholders of the financial institutions.

The value of money gradually decreases every year, while the labor that is required to earn money remains more or less constant. This has brought a new situation where people can have full time jobs, and yet be homeless at the same time. (The Conference of Mayors determined in 1996 that 19 percent of the homeless were employed.[305])

We have a system that is based upon competitiveness and scarcity. Those who are inevitably left out then need to sponge off the system for any scraps they can pick up. We tax labor to fund corporate subsidies and support the economically irrelevant members of society. This will always lead to social discontent and crime because a percentage of the population is structurally unemployed, and there is nothing they can do about it. Bernard Lietaer in his book *The Future of Money* calls this the Yang Economy. In the Yang economy, which is based solely upon commercial transactions, the needs of the community are dependent upon non-profits, labor taxation, and subsidies doled out by the commercial sector.[306] Lietaer writes, "The design of the money system is preordaining 90 percent of the investment decisions made or not made in the world. And regulations aiming at sustainability just try to correct the flaws built into our money system."[307]

Four Kinds of Wealth

When Dominators imagine wealth, most only think of money. Wealth, to a corporation, is strictly financial, and in the short term at that. This is because shareholders do not care if a corporation goes under as long as they get out and sell out first. Everything else is either going to obstruct or facilitate the accumulation of it.

Dominators don't think of trees when they are counting cash; Dominators don't think about the disadvantaged poor and homeless or families forced apart when they are checking their bank balances; Dominators don't think about the health of others, or even their own for that matter, when they are trying to suck as many dollars as they possibly can from their own labors or the labors of others.

Of course, having been indoctrinated in Domination culture, we all understand **Financial wealth** quite well. We understand it so well, most of us think it is the only kind there is. But there is also:

- **Physical wealth**—the quality of our mental and physical beings;

- **Natural wealth**—the quality of our biosphere, including all ecosystems and our air and water;

- **Social wealth**—the quality of our relationships and our cultural experience.

As we count our bills and coins, we become detached from the reality that these three other forms of wealth are crucial to the creation of financial capital, and visa versa. To be "wealthy" we need all four in balance.

For example, the Natural wealth of the planet is a one-time endowment, but our capitalistic system doesn't have a column for it on the balance sheet. That is why our system can pollute and/or destroy these valuable resources with impunity. It has no regard for the consequences because it can only see wealth in terms of the accumulation of cash. It is only concerned with Financial wealth.

Corporate culture uses society and culture (Social wealth), human beings (Physical wealth), and natural resources (Natural wealth) along with a new form of wealth (manufactured items such as machinery) to produce Financial wealth for itself. It attempts to use as little of the human capital as possible to transform as much of the natural capital as possible into as many products as it possibly can.

Our Natural wealth, which is part of what belongs to all life, is being depleted by corporate entities that merely claim these resources as its own, much as Columbus claimed the New World for Spain. It doesn't really belong to them, but they claim it anyway, move to seize it, and combat anyone who attempts to stop them using all available means. This is a system which takes nature, transforms it for profits, and then markets and sells it to consumers who use these products until they enter the next stage of their existence—the waste stream—the cost of which is borne by our civic structures. This continuous cycle of the transformation of natural resources into waste products is destroying the common Natural wealth, which is wealth that belongs to everyone.

Tribal culture, as we will examine much closer in the next section, is based upon a different set of values. Personal and cultural values will define what wealth is to any individual. It will also determine how a society is structured. To tribal people, wealth is measured in their mutually shared feelings of spiritual connection. They can feel and celebrate the sacredness of their immediate reality because they live within it, are one with it, and totally consume it as it consumes them. It fills them with an overwhelming sense of the abundance of life and complete spiritual interconnection and immersion. It is a communion in the highest sense of the word; it is a communion with life.

The Profit Imperative and the Investor's Mandate

Not so for corporations. It would be quite odd to think of a corporation as being capable of feeling immersed in lovely feelings of belonging and connection with the universe and life. It cannot, so therefore corporate culture is unable to place value on these strictly human forms of wealth—at least in the context of our current economic system which does not demand a cost be placed on Natural, Social, and Physical wealth, all of which are consumed voraciously by the corporate sector.

And so it is that nothing is more basic to the purpose of a corporation than producing a profit in the form of financial capital. All other values are secondary. If a corporation does not produce a profit, then the investors will pull out and the corporation will fail. Stockholders would be foolish to invest their money in any vehicle that is not going to give them a return on their investment. That is precisely why a corporation must do everything it can to make money, even if its actions are contrary to the principles and standards of the investors themselves.

Not only must every corporation produce a profit in order for the stockholders to continue to participate, it must continue to expand and grow. Growth is the standard that the stock market uses to determine the future prospects of corporate performance. Investors need to have faith that the profit will not only be made today, but tomorrow as well. Profit and growth come before worker satisfaction, environmental integrity, community, individual health and welfare, and principles of cooperation.

Our consumer-driven economic system measures all growth in strictly financial terms because corporate culture is the base upon which our civic structure is now placed. This means that even financial transactions that occur as the result of any degradation of human culture or natural resources are counted as contribution to the total sum of economic growth! The authors of *Natural Capitalism* gave some examples of this absurd accounting method: "Growth includes crime, emergency room charges, prison maintenance, dump fees, environmental clean-ups, the costs of lung disease, oil spills, cancer treatment, divorce, shelters for battered women, every throwaway object along every highway, and liquor sold to the homeless. When accepted economic indices so wildly diverge from reality, we are witnessing the tottering end of a belief system." I call this phenomenon "Degenerative Growth," which is a negative, not a positive, number.

Political and Governmental Control

Corporations are the most important financiers of electoral campaigns worldwide. They have access to huge amounts of discretionary income for influencing the public and the politicians who want to be elected. They have the capital to lobby for the laws that will benefit their goals, regardless of whether or not their goals may not be in the public interest. They have the power to place their own operatives within the governmental structures of the entire world. They can influence the military actions and surreptitiously command armies from corporate headquarters.

The public should be outraged that it is now common practice for corporations to actually write their own governmental regulations. When this happens, corporations can write regulations that allow them to legally pollute and endanger human life while shielding their financial interests. This is because the laws are written by lawyers representing corporations, and those laws will protect them from potential lawsuits. In effect, this legally allows them to act against the best interests of the general public. Furthermore, if what a corporation does is "legal," the public usually will not even learn of the deceptions perpetrated upon society at large. Corporations spend billions and billions of dollars every year to make sure that our world's governments satisfy their needs and wants.

Furthermore, corporate sponsored legislation can also destroy smaller scale operations by creating unfair tax advantages for the mega-corporations. Tax breaks typically favor operations that are less labor than energy and capital intensive. To add to this disadvantage, many regulations are written that are affordable for larger operations that the smaller ones cannot afford to comply with. These regulations are usually written because of the conspicuous problems that larger operations create because of their sheer size.[308] Even though a smaller company might not generate the problem that necessitated the regulation, it still must adhere to the law, putting them out of business.

And when the corporations find it more convenient, they will promote and campaign for "less government" as if the public is demanding freedom from too much government in their lives. But what we need to understand is the government should be representing our interests against the corporate interests that want to rule. Corporate entities want nothing less than total control of every facet of human society and culture. They want control over their employees, the desires of the consumer, and the natural resources that they must extract to reap their profits.

In his book *Unequal Protection* Thom Hartmann writes, "The 82 largest American corporations contributed $33,045,832 to political action committees in the year-2000 election cycle," fifteen times more than labor unions. "In 94 percent of U.S. House of Representatives races, the candidate who spent the most money won."[309] Is this the democracy envisioned by the founders of our country? Wasn't the American Revolution about freedom from oppression? I fail to see how giving power and control of our government to the rich and wealthy lives up to the ideals that were intended by the authors of the Constitution, the Bill of Rights, and the amendments which followed.

Corporations control the government, plain and simple. Corporations dictate the issues. With cash they can write the platforms of the major political parties and narrow the scope of discussion. When you add to that the fact that half of the industrial corporate wealth in the U.S. belongs to just under the top 200 corporate entities, you are left with totalitarianism—the complete control of a society by a select group. In reality, all governments are totalitarian, whether they are based on capitalism or communism. The only difference is in the methodology.

And now corporations are gaining even greater control of governments, bypassing their authority through international trade agreements such as NAFTA and the WTO. These trade agreements force all nations to adopt the most permissive and lenient environmental and labor laws of the signatory nations, under the guise of making trade fair for all parties. So if a signatory nation should choose to stop a corporation from abusing its labor force, polluting its environment, or levying taxes to prevent it from destroying local entrepreneurs, that corporation can sue for damages through secret tribunals called Dispute Resolution Panels. Public interest groups are not allowed a voice in or to witness these restricted hearings. In just about every single dispute the panels vote in favor of corporate interests.[310] These panels are not elected by the people, but are selected by the international industries themselves. They have veritable god-like powers over nations, and yet are accountable to no one. They have the power to overturn the will of the people by striking down any legislation in any country it deems appropriate to advance the corporate agenda. Under these types of international free trade agreements, democracy is an obstacle.

Corporate Rights versus Corporate Responsibility

We have the death penalty in the U.S., a device that is rare amongst western nations. It is supposedly there for revenge against someone who has committed a heinous deed against other persons. It is supposedly there to permanently remove that criminal from society. It is supposedly there to teach other citizens to think twice about committing a heinous deed themselves. Our government will kill you if the courts determine the punishment fits the crime.

But what about crimes that corporations commit? Many corporations kill people with the sale of dangerous products or unsafe working environments. Corporations break laws all the time, but all they receive is a proverbial slap on the wrist. Fines that are ridiculously low for a corporation are rarely levied and enforced.

If Union Carbide, Shell, American Tobacco, or Unocal were the names of human beings, they would be imprisoned, or even, in some of the more repressive regimes, executed. But the people who own a corporation are not personally liable for the actions of that corporation. A corporation cannot be jailed or executed. Capital punishment simply does not exist on the corporate level. A corporation can only "die" when its charter is dissolved by the stockholders or if it merges with another corporation.

But even though corporations are not required to abide by the same responsibility as private citizens, it enjoys all the rights. A corporation can sue for slander and libel. Corporations assumed First Amendment constitutional rights guaranteeing freedom of speech even though no law or court decision gave them that right. As we have covered in the Section I, it was merely a court reporter's head notes in 1886 that incorrectly classified corporations as "persons," as defined by the Fourteenth Amendment. Corporate lawyers have deceptively pointed to that definition ever since to allow corporations the same rights as a real human being.

By allowing this constitutional protection to be expanded to cover corporations, they have effectively been allowed to dominate the flow of information that reaches the public. Billions of dollars of advertising dominates the public consciousness and perception, virtually free from any regulatory checks. They have the power to say whatever they want and have the resources to see to it that their message is heard and understood by just about 100 percent of our society. This is power, not democracy.

A whole industry has been born and is flourishing as a result of the expropriation of the First Amendment. Public Relations (PR) is a multi-billion dollar industry to serve corporate interests by waging war against watchdog organizations and individuals who report on consumer safety, environmental degradation, and other exploitive aspects of the corporate world. "Greenwashing" is a word coined to describe the media campaigns to exonerate guilty corporations from their crimes in the public eye using sophisticated psychological techniques.

Corporations will go to great lengths to keep the public in a state of conflicting and confusing flow of misinformation. They fund "studies" that will contradict the truth of the consequences of their actions. They regularly create corporate funded grass roots campaigns that appear to emanate from concerned citizens, but are really funded by the corporate world. These supposed nonprofits spread false information and doctored studies by accredited professionals in the employ of corporations to confuse the public and influence lawmakers. Even legitimate information and studies that come from universities and other public sources can be twisted and molded to disseminate untruths to suit the corporate agenda.

The intent of the First Amendment was to protect personal speech in a time of history when the only media consisted of handbills and a few books. The Bill of Rights was written in 1792, and corporations did not exist then as we know them today. Surely the founders of the U.S. did not envision giant corporations monopolizing the flow of information when they wrote these basic laws to protect the rights of individual citizens.

It's Just Your Imagination

A corporation is but a concept. It is merely an agreement between individuals who band together to create a mutual business endeavor. It has a name and a legal existence, but you can't see it. Corporations exist only in our imaginations.

And while they exist only in the human mind, they have many tangible aspects that give the impression of a good sci-fi flick. Corporations can cut off pieces of themselves to birth new corporate entities, or combine with other corporations to become one. Corporations can own one another. They can be sold to a different owner. They can morph and change identities. Corporations may appear to exist as a building somewhere, but you could just remove the name and put another in their place overnight. The mega-corporations are totally inbred with each other (through investors) and share board members. And yet a single corporation will claim to be, in the eyes of the law and the Constitution of the United States, a person—but as we have seen, only as it applies to its rights, not accountability, especially for its very human beneficiaries.

Though corporate entities exist in our culture's consciousness and on paper, they have no concrete form. Though a corporation's employees have human feelings, the corporation itself cannot. A corporation might try to advertise that it has feelings, but we know that is not possible. It has no physical existence because it is a concept. It cannot feel shame or remorse over anything it does. It has no morals, so its decisions can cause suffering to others and it will have no misgivings.

Bad Neighbors

A corporation has no interest in the community it serves, except for those goals that will help it make money. Since it has no physical form, it can disappear from a community and put up shop the next day in a place that will help the bottom line. Actually, corporations quite often will abandon communities, leaving them in economic despair. They'll tell some of their employees to move if they want to keep their positions, and others they'll drop.

Many times corporations will play one community against another to get the best deal. Taxpayers fund the competition to offer absurdly generous incentives to entice corporations to come to their communities or to stay. Through subsidies, incentives, or reduced taxes, the real substance of global competition results in a struggle to endure the lowest standards as possible for worker's pay and benefits, as well as corporate responsibility to the community and the environment. And even then, after collecting their loot, they still find loopholes and move away—sometimes even to a foreign country. Many times, a corporation that vanishes also dumps the cleanup of its activities on the local economy. A community can never be stable if the businesses that offer employment have no allegiance to the welfare of its inhabitants. Communities can never be stable if citizens are constantly coming and going in search of better wages and opportunities. The cultural cohesiveness is also lost.

As the proverbial wolf masqueraded in sheep's clothing, so do corporations. Corporations would have you believe that they are a flesh and blood person. Then they use this ruse to attack people under the protections offered by the Constitution for "persons." If you use your human constitutional right to free speech, and speak against the activities of a corporation, that corporation can sue you for slander, libel, or harassment. But the corporate motivation behind it is usually to stop real humans from exercising their right to free speech, not to win a case. A corporation can outspend truly concerned citizens and force them to back off and frighten others from speaking out.

Corporations wield the power to control the flow of information and create situations that they can exploit for profit. Ocean Robbins, in his book *Choices for Our Future*, described one of the many destructive ploys greedy corporate executives have performed for their own personal profit at great expense to society:

> William Hearst, owner of a large timber company and many newspapers printed from tree-derived paper, started a campaign to stop hemp. He filled his newspapers with stories portraying hemp as a drug used by "criminals and minorities." Hearst used the Mexican slang word "marijuana" instead of "hemp" to help change its image from the all-American plant . . . Hearst was joined by Du Pont and several other companies that saw a potential for profit in outlawing hemp. For Hearst and his enormous timber company, making hemp illegal was attractive because it guaranteed an enormous market for paper made from wood. For Du Pont, making

Aspects of Domination   131

> *hemp illegal would eliminate the stiffest competition for the company's new synthetic fabrics, such as nylon.*[311]

Hemp and many other valuable resources, strategies, sciences, technologies, and methodologies that have the potential to bestow great benefit to human life are concealed from the public because the corporate potential for profit may not seem as great. This forces our society to be unjustifiably reliant on various raw materials and technologies that may harm the planet and reduce the quality of life for the profit of a select few.

Domination inherently produces prodigious waste in every sector of our society—waste that is the real and hidden costs of a system that encourages exploitation on a large scale. How does needless and uncompassionate waste correspond with values that real people have? A forest's existence, children's relationships with their parents, time spent in traffic, the manufacture and use of military equipment, litigation, the quality of a water source, the health of a person, the storage of nuclear waste, the viability of the carbon cycle—these are areas in which human beings have values that are not shared by this dominant system which caters to the wants and desires of heartless inhuman machines that devour life, and even the planet itself, for cold cash.

(I would like to note that there are many nonprofit corporations that work to protect us from and expose the wrongs performed by large publicly owned corporations. Also, most corporations are very small in nature and not capable of the heinous acts perpetrated by the mega-conglomerates that threaten us today.)

Human Cogs in the Corporate Wheel

The work ethic is one of the most critical of all the corporate values. Working long and hard increases your value and you most likely will be financially rewarded. Make sure you learn marketable skills, because if you don't you may be forced to perform menial chores in order to survive. And if your talents and inclinations don't interface with the employment opportunities at any given time, then you may be faced with unemployment. The struggle then turns to mere survival and how to put food on the table.

The Domination culture views the unemployed and unemployable as lazy good-for-nothings, degenerates, and undesirables. We look down on them even though the economy is designed to keep willing workers unemployed. This is done to keep wages depressed, which in turn increases the profit margins for the stockholders.

Employees exist to do the work to produce profit for the stockholders while those investors are free to do other things. The employees are obliged to perform and abide by the rules set by the corporation if they want to keep their jobs. If the corporation requires them to perform a task that they feel is wrong or immoral, it makes no difference to the corporation. A corporation does not care what principles an employee has as long as they suspend those standards when functioning for, speaking for, or acting on behalf of the corporation.

Employees are truly depersonalized within the framework of the corporation, even all the way to the very top. If a company executive were to attempt to act on their own personal principles, that executive would be replaced with someone who would not. All employees then are merely cogs in the great wheel of the corporate machine.

Corporations, in an attempt to streamline their operations to procure the greatest profit, reduce all jobs to make them as simple and repetitive as possible. If a job can't be automated (yet), then they want to make it an easy-to-define routine with clearly marked objectives. This allows positions and the people who fill them to be moved here or there with the least confusion. It also allows them to be scrutinized, and the productivity of each employee can then be easily quantified, measured, and compared to the others. This helps to promote competition amongst the workers. This can turn into aggression and stress when employees who want to advance and earn more must be willing to sacrifice anything to beat out their colleagues.

What effect does any of this have on the human worker? We evolved free and were created to act according to our own personal values and belief systems. What are workers to do when forced to behave contrary to their values and belief systems? Many times the choice is to betray their personal values in order to feed their families. They know that total allegiance to corporate policy is mandatory to keeping their jobs. How do people feel when they are not valued as human beings by the company, but rather mere replaceable units in the corporate structure?

Many willing and able people, most of them quite poor, find themselves looking for work, but unable to find any. How do the unemployed feel about themselves when they are confronted by the sinking feeling that they live in a society that apparently has no need for them? What happens to the dignity of an individual who comes to the conclusion that their life has no value in the reality of the world in which they live?

Entrenched

Change is necessary, and yet, no matter how bad things may seem, there are always those who fear change and will resist it. For example, I've heard it said that many of the slaves of the Confederate South were reluctant to claim their freedom because they had never experienced it.

Our cultural ways are indelibly etched into our minds, just as every human being who ever was, is, and will be. Our heritage is undeniable; our ties to our personal mother culture are absolute. Breaking with tradition is not something we human beings can do easily.

I believe that to begin moving in a new direction, we must first understand some basic principles of human culture and apply them to our current predicament. It is these principles that hold the keys to our deeper knowing—keys that exist deep within our ancient memories that can help us comprehend what it means to be a fully realized member of our species.

The beautiful thing is that the modern human will still respond to the values we evolved to enjoy, regardless of what social conditioning we may have endured. The way to these true innate human values begins with knowing what they are. Then we must incorporate them into our everyday lives, our cultural identities, and our societal structures. This will be the Vision of a new evolution.

IV

A New Global Cultural Vision

ooooooooooooooooooooooooooooo
"We are the land . . . That is the fundamental idea of Native American life: The land and the people are the same."

—*Paula Gunn Allen*

Vision Creates Reality

Every human being, institution, and culture is guided by a vision, no matter how lofty or diminished it may be. Vision, as I am using the word in this book, is defined as "the act or power of anticipating that which will or may come to be." Vision is not based upon facts, but ideas that outline and comprise the aspiration of the creation of facts to come. It is a detailed picture of potential.

Freeze-framed photographs of a flying object taken with a high-speed camera may show great detail of that object. But looking at that photograph does not reveal the direction it is going in, or even if it is moving at all.

Now suppose that the current human global meta-culture is that flying object and its direction represents our culture's vision. We cannot know where our vision is taking us if we don't know what direction we are going. The only way we can learn this is to learn our past to find out what direction we were launched from, as we did in Section I. If we can deduce that, we can clearly see where our culture is taking us and comprehend the present "freeze frame" of the now, as we did in Sections II and III. From this we can accurately predict where we are headed, learn our collective vision, and alter it—which is the purpose of Sections IV and V.

Memes are the Genes of a Culture

A meme is an idea of just about any sort that can be and is shared with other human beings through imitation and repetition. The idea of memes (a meme in its own right) is outlined in the book *The Selfish Gene*, written by Richard Dawkins, who also coined the word. The word is a hybrid of the words "gene" and "mimic."

In each biological gene there exists a small unit of information encoded that combines with other genes to create the blueprint for a biological entity. Similarly, a meme is any piece of information that is replicated, imitated, and mutually understood by any segment of a culture.

Memes have a life of their own, in many ways. They exist in our minds and only replicate themselves if they are good at replicating, not necessarily if they are good for us.[312] They replicate by using the genetic disposition of the human species to imitate one another. And of all the animal kingdom, there is no other creature that possesses this inborn skill to the proficient extent that the Homo sapiens has.[313]

Those memes that are successful are rehearsed and shared. Other helpful factors in successful meme transmission would be memes that are sensationalist, transmit warnings, help one increase social standing, to conform with a peer group, or contain references to either sex, food, or power (i.e., war or conflict etc.).

A concept, institution, and school of thought such as a science, religion, belief system, political system, nation-state, and so on, can also be called a "meme complex" or a "memeplex." A memeplex's existence is due to and borne of the smaller memes of which it is comprised. Usually the most successful memes will be those that trigger emotions or fit neatly with these culturally installed memeplexes.[314]

A memeplex contains two or more, and typically millions of memes. The reason some memes must exist within a memeplex is because of its nature, its chances for replication and survival are better, or even totally contingent upon, being bundled with other related memes within the memeplex.

This is perhaps one reason why some nonprofit groups, like environmental organizations, may have such a difficult time with their agendas. They may successfully attempt to promote or eradicate particular memes, only to end up back at square one to answer the same challenge again and again. For example, "This forest should not be destroyed" is a meme that hardly anyone could argue with. The meme it is up against, "These trees and the land of this forest exist solely for the benefit of humans," would be very weak standing alone. But the memeplex that the pro-use meme is bundled in is part of the core Domination culture value structure (which we will be examining shortly). Dominators would imagine that the forest was actually created for their personal use. Of course, this is a fatal thinking process for obvious reasons, and as a stand-alone meme it is practically laughable. But this leads me to consider that, for effecting positive change, perhaps humanity needs to focus not only on the local issues, but on the larger memeplex of Domination as well. And the root of Domination lies in its core values.

Cultural Values

The memes that contain a culture's values are the very nutrients for their cultural vision. In tribes and other forms of human societies, this cultural vision is a crystal clear view of how their life should be lived. Tribal societies know that this is *their* vision and theirs alone. The cultural vision and values of other tribes matter little to them, as long as they are allowed to live their life in accordance with their own values.

Tribal/Human/Animist Values[315]

Tribal values are the social nutrients that sustain and promote indigenous cultures. I also refer to them as Animist or Human Values—to me they mean the same basic thing. It is through mythology that these critical values, along with the way to live, are ingrained in the culture. But before we examine mythologies, we should understand the three requisite human values for a culture to properly function. If any of these preconditions are absent, the cohesiveness of the culture will wither and lose its effectiveness and relativity; the tribes and societies who rely on these critical social systems would then become vulnerable. These three core values are:

Spiritual Connectedness to the Land

First, people need to feel that who they are, what they are, the very essence of their being, belongs in the world around them. This value provides a close, caring connection to the land upon which people live. It is a sacred bond to the native peoples. This is spirituality: To feel part of the wonderment. To quote Jim Mason in *An Unnatural Order*, "Human spirituality began with awe of life on Earth. Life on Earth is the miracle, the sacred. The dynamic living world is the Creator, the First Being, the Sustainer, and the final resting place for all living beings."[316]

Spiritual Connectedness with all Tribal Members

Secondly, people need to feel that every member of their society is *equal* and having the same experience as they are. This would be the antithesis to hierarchical form that is evident in just about every aspect of modern reality. Power of decisions that govern the tribe is shared amongst all members, not certain individuals. True community values that allow people to care about one another will naturally occur when all members of that community are equals and accountable for their actions. Individuals must be responsible to function as a member and be unable to act or transact anonymously.

Spiritual Connectedness to a Common Bonding Purpose

Lastly, tribal culture will only take place in a society in which all people are connected to something that's bigger than any one of them can be individually. The connection itself, the glue that holds it all together, is the tribe or what we may think of as community. The needs of the community must be greater than the desires of the individual.

These are the memes that are the building blocks, the primary human values, which form the foundation of any human tribal culture. Our species evolved and is genetically designed to live and function in a tribal culture that espouses these three values.

Once again, the three critical values that must exist for true human culture to thrive are:

1. Spiritual Connectedness with the Land (and all life that lives upon it, as well as the cosmos);

2. Spiritual Connectedness with all Tribal Members (each individual being equal);

3. Spiritual Connectedness to a Common Bonding Purpose (providing identity and belonging).

These critical foundational values are seamlessly transferred from generation to generation through each culture's unique foundational stories and myths.

The Great Storytellers

The development of language was perhaps the first major advance in memetic transmission. Within the ancient tribal societies, the evolving and increasingly complex linguistic abilities human beings had developed gave them an evolutionary edge when gathering food or defending the tribe, as well as helped to preserve the ancient wisdom of their forefathers. Some anthropologists claim that this occurred during the period of time known as the Human Revolution, which they estimate to have begun around 40,000 years ago.

As each culture evolved, certain stories were repeated—stories of life, passion, mythical beings, and history. These stories were expounded upon and repeated again and again, generation to generation, and would form the rock upon which each separate and distinct culture would build itself upon and draw its identity. There were characters in these stories for everyone to emulate and learn a lesson from.

Most of us might imagine these mythical stories as mere entertainment, but the impact they have on us is far greater than that. They create who we think we are, individually as well as collectively. Historical accuracy is of little importance to the purpose mythology serves in human culture. The actual facts matter much less than what is the meaning, message, or the lesson to be learned.

The stories we hear as children, whether they come to us from tribal ceremony, peers, parents, books, newspapers, radio, television, or the Internet, create the foundation of our understanding of the world around us. It is a memetic narrative that describes the how and why of mutually understood and accepted assumptions that lay the foundation of a culture.

Each generation throughout the ages would pass these great stories, the essence of who they were, to the succeeding generation, and perhaps elaborate upon them as well. In this way, each and every human culture was living and evolving. Rules were not written; laws were not decreed; nothing was etched in granite. Each generation would fine-tune what was known and then add in their own discoveries by refining the stories and adapting them to keep them current and pertinent to their perception of reality. These cultures were very fluid and in a constant state of flux as conditions changed and tribal culture could and would adapt to the changing world around them.

Tribal culture was the total sum of the stories and oral traditions that were the vessels for the culture's memeplexes that were passed on and on. This is how the human naturally learns. We don't learn how to act by pure instinct. Humans evolved to learn by emulating elders and peers and sharing stories and reenacting them.

It is the sum of the tribal/societal memeplexes that is the culture for these human groupings. Culture is the mutual understanding and comprehension of reality. These memes of mythologies are transmitted through the art (including music, painting and other visual arts, and the spoken word), ceremony, and language of a culture.

Living Mythologies[317]

Human culture evolves as it is guided by what Joseph Campbell called "Living Mythologies." Mythology, especially with creation stories, serves an important purpose for any human culture. Every tribal society is founded upon a culture within which "lives" the mythology that provides the identity and social norms that live and burn in the psyche of each member. Living Mythologies serve four important cultural purposes:

Mystical Function

Joseph Campbell wrote that the first purpose of a Living Mythology is called "mystical function." Mystical function serves "to waken and maintain in the individual a sense of awe and gratitude in relation to the mystery dimension of the universe, not so that he lives in fear of it, but so that he recognizes that he participates in it."

Image of Reality

He goes on to say that "the second function . . . is to offer an image of the universe that will be in accord with the knowledge of the time."

Cultural Norms

The third function "is to validate, support, and imprint the norms of a given, specific moral order" to each member of a culture. These are the stories and oral traditions in which each individual member of a tribe would learn about life, the values, and the belief systems commonly held within the tribal unit.

Daily Guide

Lastly, the fourth function is "to guide (each member), stage by stage, in health, strength, and harmony of spirit, through the whole foreseeable course of a useful life." Once learned and incorporated, the commonly held values and social structures provide a daily guide on what is to be expected of each individual and how to behave and conduct oneself within the structure of the society.

A child growing up in a culture with a Living Mythology would feel the mystery and the awe of that child's universe, and would feel the connection and be amazed by it. As the child learns about the immediate environment, the understanding of the nature of things would be broadened and enhanced by the supporting myths. Once the child becomes an adult, the myths are not as important as are the imprinting of the social norms on that individual. Each action within the society, then, is within the commonly accepted boundaries.

It isn't the myths or the details themselves that are so important, but the moral directives the myths impart. Once learned, the growing up is over and it is time to function as an adult.

Living Mythologies are experienced. As such, they are seemingly self-evident. No texts are required for Living Mythologies. Going one step further, any written text will make any Living Mythology obsolete and useless. Times change, circumstances change, and myths and beliefs must be flexible to help teach each generation the moral principles and values that will be pertinent to the world in which that child is going to live. This is why they are "living"—they are fluid and always evolving as they gradually change with time. There was no "generation gap" in these cultures because the Living Mythologies naturally evolved with the tribal culture on a daily basis. Every generation, existing at any given point in time, felt and experienced their cultural evolution together.

Once again, those four important purposes that Living Mythologies serve are the:

1. Provision of Mystical Function;

2. Provision of an Image of Reality;

3. Imprinting of Cultural Norms (to each member of a culture through the stories and oral traditions);

4. Provision of the Daily Guide (on what is expected of each individual and how to behave and conduct oneself within the structure of the society).

Tribal Vision

The values are the foundation upon which Living Mythologies build the vision. Mythologies are the blueprint that designs the cultural vision. The cultural vision is the way to live and merge with the Universe. The "way to live" can vary quite wildly from one culture to the next.

The cultural vision of one tribe is of little consequence to another, so long as it doesn't interfere with the other culture's vision. For example, the method by which one culture selects mates or what games they play is of little consequence to another, so long as they are not intruding in an unwelcome manner.

A tribe would not care about what method another culture used to gather food so long as it did not affect their lives and culture. They would not venture into the traditional lands of another tribe without a mutual understanding. Cultural boundaries were of great importance because the land upon which each culture existed was sacred to them. It was not a commodity—it was their life.

The vision an individual or a culture holds in their imaginations would therefore be a direct consequence of the memetic values they cherish. And as we have seen, mythology is the prime memetic vehicle to present the values in a comprehendible story-like fashion, making them accessible.

Dominator Vision

Sharing these mythological memes becomes a different type of task, however, when humans are crowded together in large numbers. Humans are very good at sharing and internalizing vast quantities of memetic information. But as the global population has been increasing geometrically, the current global culture has learned, out of necessity, to become progressively proficient throughout the ages in the mass transmission and preservation of memes. Now we find that most of our memetic transmissions are under the control and supervision of various institutions, such as the corporate media or the government, and so part of the mythological makeup of our culture is created by corporate institutions.*

Overpopulation Prevents the Manifestation of the Primary Human Values

Humans evolved in tribal units called bands, which typically consisted of less than 30 individuals. As we have seen in Section I, *The "Roots" of Domination*, totalitarian agriculture has changed all that. Overpopulation begins when the size of a tribe exceeds 500 and/or a band exceeds 30 people. Overpopulation will shake the very foundation of culture, that foundation being the human tribal values themselves. Let's examine each of the three primary values individually:

Spirituality is the awe of life, and the feeling of connection to the earth and other beings, human and non-human. When people are crowded in on top of each other, our focus and energy is spent on creating space for us. We may focus on how so many other people are affecting our relationships. We may focus on how they are cramping our life style or competing with us in our careers. Our ability to fine-tune our psyche to the natural rhythms of nature is blocked or ignored. The source of our sustenance is isolated from our awareness. The land upon which we live becomes less sacred to us because we become detached from it. We seek religious experience to fill the gap, but more often than not, religion only deepens the gap between our inner selves and the Earth. And so that which is not self-evident requires Faith.

Equality is that comfortable feeling that everyone in your entire universe is totally equal to the other. This is something that very few modern people have experienced. This is because there are too many of us. With equality comes tolerance, freedom of expression, and freedom to act according to one's own values. With overpopulation these concepts are compromised. In overcrowded situations, people are forced to conform to rules and regulations set by someone else, lest the delicate balance between peace and violent chaotic societal breakdown becomes disturbed. With a large population base they are further enabled to conduct themselves in an anonymous fashion and exploit others without being recognized. Various factions and individuals will be encouraged to claim control with the ill-found belief that they have the "Righteous Destiny" to exploit, rule, and impose their value system on others.

Purpose is that good-feeling, altruistic sensation that a person may have when working together with others on something that is larger than any one of the collective individuals. But in the case of culture, I am talking about a permanent social structure, not just working with some neighbors or friends on a community project. Purpose should be as personal as it is universal for it to influence culture. The community that instills the sense of Purpose must also furnish the cultural identity for every member. If humans of varying cultures, persuasions, and inclinations are crowded together in one space, the sense of community becomes eroded. Contact with too many individuals that do not share the same cultural identity leads to conflict, especially if they are crowded together in the same physical space. Institutions are created to control the society, which is ultimately doomed because people don't really care about these faceless, emotionless, and cold entities that they work for, buy from, or are ruled by. Community relationships come to be replaced by cash transactions, and the heart of the community is replaced by institutional structures.

Dominator Values

We've examined tribal values and how they have been compromised by civilization. What might the three corresponding Dominator values be?

1. **Faith**—Sacred is the Nation (Roman Empire, U.S., etc.), the Lord, the Word of God (the Bible, Koran, etc.), the holy church, Mankind, and Salvation. The Earth is considered to be of value only in the context of how it can be utilized by humans;

2. **Righteous Destiny**—God created all life on Earth for Mankind to use at Man's discretion. God has designated Mankind to be His stewards of the planet. And some people deserve more of the benefits than other people by birthright. Our possessions and lifestyles are of more value than are the lives and cultures of other human beings;

3. **Institutional Control**—God created man in His image to rule Creation. Mankind rules through the creation of institutions (church, state, military, and corporate) that have the ordained authority to exert control over culture, society, and all life and natural systems. And as Mankind rules through the institutions, Humankind shall serve and be subservient to these man-made concepts. The value of a human life is then reduced to the level of individual productivity.

Domination culture is based upon these false, fatal values, which, as we shall soon examine, create what I term "Mythical Dysfunction."

Spirituality, Equality, and Purpose

There is no culture that can thrive without these three critical human, tribal values. We evolved with them. We are genetically designed to function within a culture that embraces them. So where does that leave us who live in an ever expanding global Domination culture? It is my contention that we are all culturally impoverished. Our global dilemma is the result of our culture programming us to:

1. **Consume and Discard**—Faith is sufficient because the planet is not sacred;

2. **Compete and Conquer**—anything can be designated as righteously ours, including the labors of other people;

3. **Separate and Self-serve**—we live sequestered lives and spend a vast majority of our lives performing tasks that have no personal meaning while we expect society and institutions to serve us.

Dysfunctional Mythologies

Domination culture is not based upon true human values, which are the preconditions to a Living Mythology. If any of these primary values are nonexistent, then the mythologies which are built upon these values will wither and become dysfunctional. We must understand why our mythologies seem so ineffective, out of date, and out of touch with the real world. We must understand this because if our myths are generally regarded as a joke and a lie, the four functions that myths provide will malfunction because no one accepts the foundation they are built on.

Our religious texts, most notably the Holy Bible, are texts that speak of man's superiority and of the Chosen. These contain ancient stories and myths of another people in another time that are no longer valid in today's global culture. These are mythologies that are as relevant as Zeus or Mithra. These are dead mythologies. They are dead because they are written.

If mythologies do not "live" in the sense that they continually evolve, the four functions that Living Mythologies provide become dysfunctional and no longer serve the interests of human beings and human culture as much as they do the institutions of authority that create and impose them artificially.

We've closely examined Campbell's four functions: Mystical Function, Image of Reality, Cultural Norms, and Daily Guide. When applied to today's Domination culture, these functions become warped and distorted. Let's look at each function:

- **Mystical Function** is not rooted in the world around us, but in artificial concepts that serve to separate and control.

- The **Image of the Universe**, which is written and espoused by those institutions that are entrusted with providing it, does not correlate with the body of scientific knowledge that is generally accepted as a more accurate rendering of reality. Our knowledge and our foundational myths are completely at odds with one another.

- The **Imprinting of Cultural and Social Norms** emanates from religious, corporate, political, and government institutions with an agenda to control. Although parents or tribal leaders of a besieged indigenous culture may try to teach different values than the Dominators, the power yielded by the institutions is too powerful and omnipresent. The power can be intentional and violent, or subtle and psychological. And it never stops. Today's 4,000-year-old monotheistic religions, guided by ancient texts, hold their adherents back in the learning stage, stuck and mired in the muck of details that have no use in the modern world. We are continually doused with religious dogma and propaganda. We are forever lost in this third stage of imprinting of the norms, and not free to evolve and create our own cultural ideas. We are, in essence, kept en masse as children by our institutions.

- The **Daily Guide** becomes a farce, merely an impersonal set of institutional rules, laws, and regulations set up to control and keep the population working for goals that have no real personal meaning.

The Foundational Rule

Crowding humans on top of each other will eventually erode culture because it is against human nature. We've examined the effect of the advancement of technology on the quality of human existence. The quality of our food, water, and air has been insulted. Humans have become detached from the natural world. We are mentally drifting into prearranged realities lacking in the full range of human sensory aptitudes. We are not fully realized as human beings.

Overpopulation decreases the quality of life.

History and science bear this out: Every society that created unsustainable population levels throughout time has failed or experienced drastic decline in all areas of life: health, crime, mental health, environmental health, social equality, and moral and ethical responsibility. This is an historical truism and what I call the "Foundational Rule." This is the root of our dilemma.

Headed to Extinction

Any species presented with an abundance of food will increase their population, whether it's mice or men. It's the law of natural selection working, without knowing what the real reason is. All "Mother Nature" can see is that this particular organism (in this case, the human species) has developed a niche, giving it access to an unlimited supply of food. It has defeated all foes and has no predators. It will be rewarded with more and more of its kind, until the natural limits are reached.

The only problem here is that the type of limits the human race is going to be coming up against will not result in a gradual tapering off of our growth rate. If we don't soon make these corrections ourselves, they will be made for us. Nature will correct the imbalance all by itself, against our will. The imbalance we created was foisted upon the natural world so quickly it has not had time to change in response. Nature has a way of moving slowly. But let us not delude ourselves into thinking that just because we can drive to the grocery store today and purchase all the food we can possibly eat, that we will be able to do that fifteen years from now. The immune system of the planet is beginning to rebel against the human species, and since the imbalance we have created is so enormous, the correction is bound to be as deep and decisive.

Every day we add 250,000 people to the planet. That's almost 100 million each year. This requires an additional 12.5 million acres of new fertile land that must be put into production every year. The problem with these figures is obvious—the surface of the planet is a finite measurement. In order to provide food and water for this rate of growth, we need more land than is available. People are starving to death in greater numbers each year because our agricultural system is not capable of this expansion. There is no technology to support our burgeoning numbers. And even if there was, it would only delay the inevitable conclusion that somewhere along the line, humans must limit and decrease population levels on a global basis. If we do not limit our numbers, Nature will do it for us in the most frightening and catastrophic manner—through starvation and disease.

Obviously we, as a culture, must address the population explosion on a global level if the human race is to survive. I believe we must decrease our global population before we exhaust our resources. We simply can't keep adding billions of humans to the planet.

Radical Change (of Direction)

Going back to the gathering/foraging tribal lifestyle we evolved to enjoy is not going to happen. We certainly are not going to move into caves or mud huts as our ancestors did. It is simply impossible. It is difficult to imagine tearing down cities and replacing them with forests where people could harvest their own wild-grown food. I realize that this is patently absurd.

But this is where the concept of vision needs close contemplation. In what direction are we going? Is the proverbial cultural flying object (which I mentioned at the start of this section) headed towards peace and tranquility on Earth, or are we headed towards a brick wall? Are we nurturing natural life and creating peace, or are we destroying the planet and murdering each other? It seems to me that the Dominator vision will eventually destroy the bio-system upon which all life on Earth depends.

We (meaning the members of the Domination culture which consists of most of the world's current population) need to regain that which we have lost—but not in the context of the natural habitat for wild (indigenous) humans. That paradise has been long gone for quite some time. We need to create a new paradise, and at the same time regain and sustain our very humanness, along with the needs and natural tendencies of the human creature. But we must now accomplish this in the context of a technological, computer-driven reality that includes global information-sharing. We must also foster a mutual global understanding of environmental protections and diversified cultural evolution.

But once again, this cannot be achieved overnight. Culture and "psychic mobility"[318] moves at a snail's pace. The concept that societies can change without violence is predicated on the innate human capability of creative evolution and flexibility. The mass mind *is* capable of change, but it moves slowly. And as we will see next, the mass mind is indeed changing at a pace most of us are unaware of.

Three Subcultures[319]

There are three subcultures of the Domination culture that are currently cohabiting the planet (excluding the extremely small number of scattered indigenous animists that are on the verge of extinction). Let's take a look at each one, and then we will examine a trend that is rolling underneath the cultural fabric of the world like a tsunami wave headed towards the shores of our collective future.

Modernists

The subculture of the Modernists is currently (as of 2003) the largest group. They are the secular product of the Enlightenment, and share many values with the Traditionalists, with some exceptions. They created the Industrial Revolution, the results of which we are contending with today. To them, God did not necessarily actually give man the Earth to rule, although many of them may profess to this belief. But they do believe that mankind evolved to be the superior species, and deserves control of all life and the natural world. The beliefs and perceptions of reality of the modernists is what we see reflected by the corporations, media, and nation-states of the world, with perhaps the exception of those nation-states which are controlled by the next group, Traditionalists.

Modernists can include such diverse groups as liberals and conservatives, rich and poor, from the banks to the churches, from the military general to the stock boy at the corner grocery store, from a pimp in a seedy bar to a Supreme Court Justice. They believe in the system, that there could not possibly be another way to live but in our civilization as we know it to be. There are four categories of Moderns:

- Upper-class Business Conservatives—wealthy, well dressed, and in control;

- Conventionals—career and family orientated, but disgruntled, cynical, and not interested in public issues;

- The Striving Centers—work hard to "make it" and be a success, and yet find themselves falling behind;

- Alienated Moderns—angry, lower class, and undereducated who completely reject the values and worldviews of everybody else.

Traditionalists

Traditionalism is a countercultural movement that has its roots in 1870. Traditionalists believed, as they still do, that whatever the current cultural fad happens to be is a deviation from the values of a better time. They yearn for times past when they imagine that life was simpler and easier. They long for what they imagine were the principles of a more decent, honorable, moral, ethical, and just time when authority was trustworthy and neighbors in communities helped each other. They distrust Big Business. They trust the Bible or the old American Ways (or the old Islamic Ways, or the old German Ways). They fantasize of the purity of a past that seemed to them simpler and less complicated. They are the true Believers, afraid of the rapidly changing and fickle trends the Moderns keep imposing upon them. Much of what they imagine of the "good old days" is conjured by their own impressions of how things were at another better time. Included in these numbers are groups such as the "Religious Right," the epitome of Traditionalism. They can also include other backward-looking groups such as the Mormons, the Christian Coalition, Islamic Fundamentalists, the Nazis of Hitler's Germany, evangelical Protestants, and the Ku Klux Klan.

Cultural Creatives

This last group known as the Cultural Creatives (CCs) is the newest, having its roots in the 1960s. They have no real organization that distinguishes or defines them. They are split between social and consciousness movements, each branch having many sub-branches, making them extremely diverse. They are viewed as being marginal, existing on the fringe of society. CCs are quite often ridiculed by Traditionalists and especially the Modernists. They are mocked and labeled as being "Tree-huggers," "New Agers," "Feminists," "Liberals," or "Animal Lovers," to name a few.

The group has two types split almost 50-50. The Core group are the creative leading edge and are activists, creative artists, writers, and in fields such as psychology or alternative health. The other group, known as the Greens, are a bit more conventional, less active, following the viewpoints of the Cores.

CCs tend to define wealth in other terms than the Modernists who think of wealth in mostly financial terms. This is why CCs, as a subculture, may not coalesce as a group as easily as the Modernists or the Traditionalists. The Modernists can rally around the value of financial wealth and Dominator/Corporate values, while the Traditionalists primarily rally around images in their minds of the golden days of yore and salvation of their souls. These easily identifiable concepts are simple and coarse, making the adherents of both subcultures easier to identify.

That is certainly not to say that CCs don't see financial abundance as wealth. CCs simply have the additional capacity to value and recognize the inherent wealth in the other three modes we covered in Section III: Physical, Natural, and Social. (And although most Traditionalists may believe that Salvation is a more worthy pursuit than the lust for financial wealth, they most likely still feel that financial prosperity is the most important form of Earthly wealth, nonetheless. The lines of demarcation are grey, but distinct nonetheless.)

Cultural Creatives Unite!

What has become obvious to me in the course of researching this book is that the various members of the subgroups of the Cultural Creatives must learn to recognize each other and support all causes that fall within their realm. Up until now, our many hats have obscured us from each other. The good news is a point that I hope this book, *Cultural Vision*, has made abundantly clear: There is great *strength in diversity*, and we need to build on that essential truth.

It is also true that CCs may be strewn about in a variety of issues and causes, and they may belong to unique groups where attention is focused on activities and topics peculiar to their individual interests and concerns. But one thing that unites all CCs is that they routinely question the basic assumptions that hark at the root of Domination, whether they are Traditionalist or Modernist in spirit.

This is not about divisive action to create yet another "Us versus Them." Cultural Creatives, as a group, would encourage Modernists and Traditionalists to seek their own fulfillment, just as tribal societies have little concern with what the neighboring tribes believe or practice, so long as they are not intruded upon. Let the Modernists have their cash. Let the Traditionalists yearn for, or even create their own version of reality of how they want to live. But there must be room for the Cultural Creatives to practice their beliefs. And there must be a consensus to allow for cultural diversity and ecological sustainability.

And where and when beliefs collide, in our currently overpopulated global situation, compromise needs to happen. But, let's face it, some things cannot be compromised, such as the wholesale destruction of the biosphere, rampant disease and degeneration, poverty, homelessness, and genocide to name but a few. These symptoms of Domination are not acceptable. The causes of these wholesale global degenerative events must be terminated. This is not negotiable because activities that devastate the Earth affect everybody. Human cultures cannot evolve or sustain themselves on a ruined planet or a toxic biosphere, and no human entity has the right to impose this curse on another culture. Many Traditionalists and some Modernists would also agree with this environmental imperative in keeping our planet healthy.

As we learn to cooperate and repair our fractured relationships with the planet and each other, Traditionalists need to understand that they cannot impose their sense of morality, their sense of right and wrong upon other groups. In our society this well-established but sometimes disregarded concept is termed "Separation of Church and State."

Likewise, the Modernists need to understand that their ability to live meaningful lives, prosper, and even generate large sums of cash is going to be vastly enhanced in a world with environmental integrity that is not falling apart at the seams. Everyone has the duty to protect our planet and keep our biosphere healthy and viable.

Trends[320]

In 1965, the Modernists numbered about seventy million people and comprised about half of the American population, but now they have dropped, percentage-wise, to about 47 percent as of 1999, numbering ninety-three million.

The Traditionalists are dropping much faster, from a little under half of the U.S. population in 1965 (numbering at about sixty-five million) down to merely 28 percent today (about fifty-five million).

The Cultural Creatives are the fastest growing subculture today. In the 1960s, the CCs comprised only 5 percent of the U.S. population (numbering about five million). Now that percentage has climbed to 25 percent (numbering at over fifty million) and is continuing to grow. Consumer choices are increasingly based on social motives, even though these products may cost a little more than their more exploitive counterparts. Over a quarter of all Americans are basing their spending choices on these values.[321] More and more investors are building socially-conscious portfolios, giving some incentive for corporations to give at least some consideration to the social and environmental impacts they may have. Furthermore, over the past thirty years the social and the consciousness movements, which the CCs consist of, have been experiencing a convergence and becoming less distinct and more inclusive. I believe this trend is interconnecting and strengthening the individual beliefs and values of all the movements as they discover the common ground that they all share.

Furthermore, this trend is not just here in the U.S.; it is reaching around the world. From Europe to Asia to South America people are realizing the dead-end path Domination is taking us. Citizens of the world are networking and learning what needs to be done. With the horizontal proliferation of networking available with the advent of the Internet, the Cultural Creatives of the world are likewise increasingly able to cross-connect over a variety of issues and causes as well as over borders and continents that formerly kept them apart, forging new relationships, partnerships, lines of communication, and cooperation.

The Memes of Civilization

So we can see that the mass mind is changing. Let's take a closer look at the specifics of these changes by understanding a few critical memes that compose it. The memes that created and supported the civilization of the Totalitarian Agriculturists are firmly fixated in the Domination culture memeplex. To most of us modern day human members of the global culture, these memes are almost universally accepted as self-evident truth. It would be difficult for us to imagine any other way to be.

Daniel Quinn, in his book *Beyond Civilization*, and George B. Leonard in *The Transformation*, have identified some of these memes. I have included a few of theirs along with a few of my own in the list below:

- The world was created exclusively for the human species to exploit and mold to fulfill its own short-term goals and needs.

- Growth is an indication of prosperity.

- Having a class of people known as farmers grow and process your food is much better than doing it yourself.

- We must all engage in procreation to create our families.

- Everyone should have a job that pays them for their time and effort in order to pay for housing and food.

- Competition is healthy and desirable.

- Our culture represents all of humanity.

- Men are supposed to be leaders—hard, tough, aggressive, conceptualizing and in control while women should be followers—nurturing, soft, docile, obeying, and intuitive.

- Ours is the one right way to live and everyone should live like us.

- Civilization is humanity's ultimate invention and can never be surpassed.

- Civilization must continue at any cost and must not be abandoned under any circumstance.

Even though these memes would be recognized immediately by members of our culture as Truth, they would not be readily understood or comprehended by a member of a tribal culture that had no previous contact with our way of life. You can go anywhere in the world and find that all of these memes are present wherever food is under lock and key. These memes are globally accepted tenets because the Domination culture now encompasses 99.99 percent of the global human species. Although, as many people are only too quick to point out, there are many subcultures that seem to us to be radically different, in reality the similarities are far more numerous from a memetic cultural standpoint.

Replacement Memes for a New Vision

If the Domination culture could be memetically altered, we may be enabled to evolve into healthier, more sustainable, humane, and freer cultures. Let's look at the Dominator memes already identified, and then I'll offer alternate memes to consider which will illustrate how they might be changed.

The world was created exclusively for the human species to exploit and mold to fulfill its own short-term goals and needs.

—or—

Humans belong to the planet Earth and should always take care not to alter or destroy bio-systems upon which the web of life is dependent.

◆ ◆ ◆

Growth is an indication of prosperity.

—or—

Sustaining and steadily maintaining our populations and businesses is an indication of environmental integrity and long-term prosperity.

154   Cultural Vision

◆   ◆   ◆

Having a class of people known as farmers grow and process your food is much better than doing it yourself.

—or—

Harvesting wild foods and locally grown produce with the cooperation of community members in which all members play a role is healthier and more sustainable.

◆   ◆   ◆

We must all engage in procreation to create our families.

—or—

Lowering our population base or at least in balance with the local environment is critical to our long-term survival.

◆   ◆   ◆

Everyone should have a job that pays them for their time and effort in order to pay for housing and food.

—or—

There are many peaceful ways for everyone to collectively help with the production and distribution of locally grown food. All people deserve shelter and health care.

◆   ◆   ◆

Competition is healthy and desirable.

—or—

Cooperation ensures that no one need lose so that some may win—cooperation builds community and fosters human values.

◆ ◆ ◆

Our culture represents all of humanity.

—or—

All cultures deserve the right to self-determination.

◆ ◆ ◆

Men are supposed to be leaders—hard, tough, aggressive, conceptualizing and in control while women should be followers—nurturing, soft, docile, obeying, and intuitive.

—or—

Both sexes have natural capabilities in a broad range of modes and leadership roles. Forcing individuals to adhere to prescribed popular notions of what it is to be a man or a woman limits personal and societal development and possibilities.

◆ ◆ ◆

Ours is the one right way to live and everyone should live like us.

—or—

There is no one right way for humans to live, as long as they do not intrude upon anyone else's way to live.

Civilization is humanity's ultimate invention and can never be surpassed.

—or—

There are many peaceful ways to live in harmony with the world and the people around us.

◆ ◆ ◆

Civilization must continue at any cost and must not be abandoned under any circumstance.

—or—

If one way doesn't work, together people can invent new methods and new societies that better serve the interests of humans and all life.

Four Categories of Human Experience

Of course, memes need to be organized so that when they are attempting to replicate they won't mutate or become compromised. Corresponding with the four kinds of wealth we examined in Section III, *Aspects of Domination*, are four categories of human experience. As you may recall these four kinds of wealth are Physical, Natural, Social, and Financial.

Correspondingly, I find there to be four primary, and sometimes overlapping, categories of human experience:

1. Health,
2. Environment,
3. Relationships,
4. Commerce and Government.

There really are no others. All human experience will fit into one of these four categories.

The rest of this section is organized around these four categories. In Section V, *The Cultural Evolution*, we will dissect and break down each category into smaller and more specific memes. But first, let's get the overall view and establish some basic precepts of this new memeplex based upon the three primary human values before we zero in on details.

HEALTH

The birthright of every human and all the other creatures of the Earth is to enjoy a full life, filled with vitality and free from debilitating diseases and degenerative conditions. Yet it seems that diseases of all kinds—viral, bacterial, and degenerative—are epidemic throughout our planet. Our well-funded medical science institutions always seem to be on the verge of great cures and discoveries, yet exotic new diseases as well as old ones continue to crop up all over the world.

To be physically healthy is to live without excessive degeneration and diseases. To be mentally happy is to feel spiritually connected. The medical and psychiatric establishment is overwhelmed with newly created diseases and degenerative conditions, but they can only do so much. The medical physician schooled in the western allopathic mode of thought typically only tries to combat the manifestations of disease, not the root cause. And in the majority of cases, these painful conditions are the result of poor eating habits. All alternative modes of preventive health practices in balance with traditional health care systems must be made available to all human beings.

If human beings want to stay free of disease, they need to examine the diet of our ancestors. Whole foods. Wild foods. Unprocessed raw foods cooked as little as possible. They would need to eliminate simple sugars, chemicals, and all animal products and center their diet around grains, fruits, vegetables, nuts, legumes, and seeds.

To be healthy is not only avoiding disease, it is also living in such a way that fosters vitality, happiness, and serenity. The level of health, both mental and physical, determines the quality of life. Being sickly reduces the quality of the present moment. When you come from a place of positive emotions, attitudes, and vitality, you can create your own destiny. Every human being deserves this.

ENVIRONMENT

The biosphere is the sphere of life that extends from a few feet below the surface to a few hundred feet up into the atmosphere. More than just the world around us, the biosphere is our life-support. More than just our life-support, it is our Mother. More than just our Mother, it is the Creator of all life. It is a delicately balanced, closed system that will react in ways unknown to any unnatural disturbance. Yet the Domination culture seems to care so little about the ecosystem that produced us, as well as all the precious life forms we share this corner of the Universe with.

To live in an environmentally sustainable way would require us to stop our ritual degradation of the planetary ecosystem. To live in an environmentally sustainable way would require us to foster the basis of all life on earth. On a global level, we need to do much more than just stop our foolish annihilation of life; we need to replenish it. We need to allow all life, especially non-human life, to thrive (we depend on it). We must nourish our ecosystem, which in turn is responsible for our own nourishment.

Nature, in all her abundance, has provided humankind with all matters of provision for all of time that our species has been in existence. At the beginning of Section I we noted the symbiotic nature with which all life has evolved to depend upon and coexist with all manners of the rest of creation in varying and many times undetectable ways—and of which our species is also a component, unbeknownst to most of our culture. And so it is that an insult in one sector can precipitate injury in another. After a profusion of attacks on individual species that are integrated into our ecosystem upon which we depend, the integrity of the system as a whole can be called into question, and with it the survivability of our own species.

Three strategies must be applied for the ultimate survival of Earth and all life. The late David Brower calls this "CPR."[322]

- **Conservation** is to utilize our planet's natural resources as little as possible.

- **Preservation** is to be certain that all ecosystems are left intact and fully functional in their natural state.

- **Restoration** is to take that which we have spoiled and return each altered segment to its original pristine condition.

RELATIONSHIPS

Cooperation (in place of Competition) is the value for determining how human relationships can help us lead meaningful lives. Even our relationships with Nature, other species, and the unseen powers could benefit from shifting to a foundation built on cooperation. The ruthless competition in our culture is flaunted at us as healthy and good for our spirit. While a little competition may be good at times, as a basis for all relationships it can prove harmful. Why are we fighting each other on so many levels all over the world? What is blocking us from achieving true peace?

To create new cultures based on cooperation, we first need to understand how to identify and eliminate the factors in our culture that breed crime and violence. We need to learn how to foster cultural diversity, human expression, and local autonomy. We need to teach life and those values that support it. Artistic expression, effective communication skills, and cooperative techniques are some of the abilities that will build a stronger culture, and a more peaceful and prosperous society. The localized community that is self-sufficient and sustainable will be more caring towards its citizens.

To create new cultures we must encourage free expression and local cultural evolution, free of global and corporate influence. As Gary Paul Nabham wrote in *Cultures of Habitat*, "To restore any place, we must also re-story it, to make it the lesson of our legends, festivals, and seasonal rites. Story is the way we encode deep-seated values within our culture. Ritual is the way we enact them . . . By replenishing the land with our stories, we let the wild voices around us guide the restoration work we do. The stories will outlast us. When such voices are firmly rooted, the floods of modern technological change . . . won't have a chance to dislodge them from this earth."[323]

COMMERCE AND GOVERNMENT

Our governments supposedly exist to serve the public, to protect us against harm and guarantee basic freedoms. These globally universal freedoms include the right to creative expression, to pursue happiness and fulfillment in all aspects of life, and to practice personally meaningful systems of belief without interference from any authority. Our businesses are private institutions that supposedly exist to serve the needs of the people. And yet, as governments and businesses conspire to benefit each other, millions of people suffer the anguish of destitution in a world of abundance. People resort to back stabbing and dirty politics in the workplace to secure better positions on the corporate ladder. With all the work that needs to be done to help the poor and the disadvantaged, to restore and protect the environment, there should be no unemployment. Why do we compete with each other for the purpose of defeating perceived competitors when we could gain so much more by cooperating with each other?

To create a business and political environment that will best serve the interests of the people, we must halt corporate and governmental degradation of human culture. Corporate entities must be prevented from mass consciousness programming and political control. Business deals must not create human desires for the business community to exploit, but identify human needs to fulfill so business can serve, not rule.

To create a business and political environment that will best serve the interests of the people, we must foster community, cultural, environmental, and social responsibility in all business transactions and government programs. Elected public officials must be selected by accurately distilling the true desires of the public that they are obligated to serve and whose interests they are mandated to solely represent. To further those goals, corporations must exist to serve governments, societies, and communities only when the primary goal is for the betterment of the general human population and all humanity by providing products and services that are life-enhancing.

We must redefine wealth and the roles both governments and corporations play in our lives. By creating new local and global currencies we can stabilize local economies and empower local communities to self-determine their purpose and identity. There are complimentary systems of currency that can coexist with national currencies, which would create new wealth and greater security for all people. We should be taxing consumption, waste, pollution, and environmental degradation as we shift our tax base away from labor. Taxing labor only serves to make human work more expensive and jobs scarce. This would also shift the burden of cleaning up waste from civic structures to the corporations and consumers that create it. This would bring about a new evolution in how we do business by moving towards a service economy that would naturally recycle goods, reduce raw resource extraction and waste stream volume, and restore those parts of the planet that have already been degraded.

Corporate Responsibility

The corporate model is the current primary model for modern society. Its values are the epitome of the Domination culture. Its memes square with each primary Dominator meme. Compare them and see this is an obvious truism.

It is the corporations that have the power to influence, control, and ultimately demolish culture. By applying the four categories of human experience, we can see that corporations can only degrade the quality of life, particularly if they are allowed to function under the false presumption of personhood. It's inherent in corporate structure. Let's look at each one:

- **Corporations promote disease.** Corporate entities the world over will push products and foods through advertising propaganda onto the populace with no regard as to the ill effects those products may incur on people.

- **Corporations promote environmental degradation.** Corporate entities the world over are unceasing in their voracious appetite for the natural resources that come from the earth and will deliver as much environmental pollutants to our air, land, and water as governmental restrictions will allow.

- **Corporations promote violence.** Corporate entities the world over will always make a profit keeping their employees in tight competition with each other and providing technology, services, and products to militaristic organizations worldwide.

- **Corporations promote involuntary servitude.** Corporate entities the world over will always promote an economic system that allows them to extract as much profit out of their workers as governmental restrictions will allow.

Applying the Memes

Nobody wants the world to be riddled with disease, pollution, violence, and involuntary servitude. Our institutions may cause these factors, but institutions are not human beings. Institutions are not interested or swayed by the meme pool of culture, except for how it can be exploited or altered for the exclusive benefit of the corporation.

Of course, this madness must stop if we are to go on. We have to change the direction in which we are headed. We must reduce our numbers and alter our habits. We need to abruptly cease many of the things we are currently doing (i.e., poisoning, killing, enslaving), while fostering life-enhancing practices (i.e., cooperation, diversity, nourishing) to replace those things. The Domination culture has domesticated the human race itself, along with countless other species of animals, and the rest of the natural world. We need to learn how to be human once more.

So, in the next and final section will come a set of memes of a more specific nature. These memes will reflect a more human and humane global societal infrastructure which future generations can model and use as a blueprint for the creation of new cultures.

V

The Cultural Evolution

> Evolution: "1. b (2): a process of gradual and relatively peaceful social, political, and economic advance or amelioration often contrasted with revolution."
>
> —Merriam-Webster's Unabridged Dictionary

The Cultural Evolution

Each meme in this section is part of the memeplex I am calling the "Cultural Evolution." It is not a revolution in the sense that people are going to rise up and battle the establishment, governments, and institutions with guns and grenades. This is not about creating a centralized authoritarian government. This is about less governmental and corporate influence, more diversity and freedom, coupled with a continuous movement in the direction of decentralization. This is about sharing with one another a new direction for humanity based on cooperation and true human values, which every human being is genetically inclined to embrace.

The Cultural Evolution is about true cultural freedom—the freedom to live within a global and local society that embraces human values, pursuit of happiness, and cultural diversity. After all, we are human beings, not consumers or employees, and we all own this birthright. As we have established, the values which can be considered human are:

1. Spiritual connectedness with—and sanctity of—all life, land, air, water, natural resources, and the planet Earth;

2. Spiritual connectedness with people, and the assumption of the human rights of equality and self-determination;

3. Spiritual connectedness to the local community through regional and/or local identity and purpose.

The Evolution is Underway

What is so wonderful about the memeplex of the Cultural Evolution is that the rapidly expanding subculture of Cultural Creatives will embrace it in its totality. The groundwork for the changes that will take place is already being put in place.

These are the steps we must take together, as a global species, as a global culture, as a global society. I have loosely organized these alternate/replacement memes in the four categories of human experience: Health, Environment, Relationships, and Commerce and Government. Many of these ideas could certainly overlap into other categories, but I've placed them in the one that seemed most relevant.

Each of the following memes belongs to the memeplex of the total vision of the Cultural Evolution. They are not parts and pieces that we can hope to tinker with one at a time, but rather must be taken as a whole. Their ability to survive within a foreign memeplex is compromised by the fact that they are fashioned from an entirely different set of values. So a memeplex requires the presence of each of its constituents. The memeplex and the memes depend on it.

◆ ◆ ◆

HEALTH

Public Education on Nutrition and Exercise

The true facts of what a healthy diet is must be made available to all humans throughout the world. This information must come from non-corporate sources that have no consumerist agenda. Corporations should have zero input in any forum on human nutrition. To keep insurance rates low, reduce human suffering, and encourage vitality and well-being in our culture, we need to promote health by dispensing the appropriate information.

Guaranteed Health Care for Everyone

Today, one segment of our society will get a CAT scan for a simple headache, while others do not have access to basic prenatal care. The majority of U.S. citizens, for example, have no health insurance because they can't afford it. There are many families who go bankrupt as the result of costly medical procedures for which they are not covered. This is simply wrong.

All humans throughout the planet deserve to be treated humanely, and this means tending to their basic health care and education. Each one of us, along with every human being born into this world, deserve nothing less than Universal Health Care. The community, the local government and/or the national government should share the cost.

Biodiversify Food Crops

Today the genetically identical machine-harvested plants are prone to disease due to their identical cloned nature. By keeping the same crop in a field season after season, the pests and diseases to which those plants are vulnerable have a cornucopia of sustenance to live on that no chemical could dissuade.

There remain many seed banks that contain thousands upon thousands of diverse plants species and varieties that could be used to protect our future from devastation by blight or famine. If they go unused they will eventually spoil, and so it is critical that these resources are cared for. Variety is not only the spice of life; it is the critical key to the survival of the human species.

Halt Chemical Treatment and Genetic Modification of the Food Supply

Chemicals are liberally sprayed on food and feed crops that wreak havoc on the human immune system and hormone receptors that regulate the internal glandular function for humans as well as other animals. Many of these chemicals kill life outright and are sterilizing the soils of our planet, making natural life impossible.

We need to immediately halt all chemical spraying of crops. There are better and more effective methods of pest control that have been used successfully for thousands of years that do not require chemicals. The use of chemicals is a tactic by large corporations to secure large short-term profits with no regard to the health of the people.

Other dangerous food technologies threaten to create havoc with our sustenance as well. Irradiation and genetic modification must be stopped now. There are many agricultural techniques that exist now that can naturally provide us with nutritional food without altering the very basis of life by tinkering with snips of DNA.

John Robbins devotes several sections of his book, *The Food Revolution*, to this issue. Many people would consider a stance like this to be anti-technology, and that is not the case. The fact of the matter is that scientists working on introducing the mass production of genetically altered organisms into the biosphere are doing it at the behest of corporations that are only seeking short-term financial gain, which is the primary goal of any publicly held corporation.

These careless actions threaten our entire food supply, risk widespread crop failure, have unknown side effects, have not been tested for long-term safety, can mutate and produce toxins, can produce unforeseen allergic reactions, have a decreased nutritional value, could produce antibiotic resistant bacteria, and encourage increased use of herbicides.

We must be clear—no genetic modification is necessary. Furthermore, all the healthy diverse genetic information is available using methods that have been proven throughout the years and are already stored in seed banks throughout the world. This is the promise of our future.

Vastly Increase Soil and Food Quality

Food for human consumption is presently grown on land that is bereft of the minerals necessary for healthy plants, as well as healthy bodies. This is because the large monolithic corporate agribusiness complex cares not one iota about the nutritional quality of our food. Money and financial growth are the bottom line.

Farming is a most sacred work. We need local farmers to care for the land and keep it vital. To increase the quality of our food, we need to return to organic farming techniques and small-scale farms that respect the land they are entrusted with. There are many methods to revitalize our soils, such as with volcanic rock dust and specialized algae, as well as proven natural farming techniques such as crop rotation and allowing fields to lie fallow between harvests.

Integrate Holistic and Allopathic Modes into Health Care Systems

We go to a medical internist every time we get a cold or the flu, who in turn writes a prescription for medication. We take drugs to mask symptoms, forcing pathogens to manifest themselves in different forms. Additional drugs are often necessary to control diseases that keep popping up in new places, or continually return. Sometimes, in life threatening situations, strong medicinal drugs have their place. Many times, however, drugs can exasperate the problem, making it worse by weakening the immune system.

We need to be able to employ all disciplines of health training to ensure we've looked at every angle of why a health challenge exists before any decisions are made with how to handle the problem. There are several healing disciplines and techniques that boost your body's own natural immunity to fight off disease. True healing emanates from within.

Integrate Mind/Body Disciplines into Educational and Health Care Systems

Personal stress and domestic mental dysfunction is rampant in our culture. The poorer classes are angered and stressed for basic physical survival. Psychiatrists and personal analysts are the rage among the well-to-do, who have their own set of mental traumas to deal with in our inhuman societal structure. People turn to addictions to find comfort. What many people really need is access to mental disciplines that will help them expand their consciousnesses rather than become repressed by hierarchy or stressed by predicament.

The personal benefits of such well-known disciplines as yoga, meditation, and regular exercise are tremendous and universal. People need to learn how to tap into their inner resources and become connected with their minds, bodies, and environments which would help them cope with their lives, become more motivated, and act in more socially responsible ways. The benefits to society at large would be great as more people became more in touch and in control with their own minds and bodies, creating a more peaceful and directed society that would have both renewed purpose and lowered domestic violence.

ENVIRONMENT

Increase Taxes on all Polluting Industry—NO Loopholes

We reward polluters by using public money to clean up a mess created by industry, while industry pockets the profit. If a group is involved with destroying land or polluting our collective environment, they should pay heavy taxes on profits and have the capital up front to repair any damage incurred before the damage is done. We must tax industries and practices that degrade and insult natural resources. We must drastically cut carbon dioxide emissions. One good place to start might be the billion-dollar subsidies to petrochemical conglomerates and take that money, investing it in sustainable, renewable energy sources.

Products and business practices will be designed so as to mimic natural processes that operate in a cyclical rather than linear fashion. In this way discarded products become raw materials for new products rather than decaying toxic matter that is decomposing or leaking in landfills.

Vastly Increase Funding for Climatological and Dynamic Oceanographic Research

The warming trend we are witnessing has the potential for triggering a global ice age by impacting the Gulf Stream's flow north delivering warm water. This must be closely monitored and solutions hammered out.

Among some of the solutions that have already been proposed are cloud seeding to avoid precipitation in the Greenland or Labrador Seas; barge loads of evaporation-enhancing surfactants; or bombing ice dams to prevent a buildup of fresh water.[324] The warm water conveyor we have in the Gulf Stream must never be allowed to fail, because that would be the end of our world as we know it. And because we know that it will happen someday, it is all the more crucial that we prepare our world for such an eventuality.

Halt all Deforestation

Trees and entire forests are clear-cut every day, leaving no life behind. This is being done in large part for meat production, lumber, and paper products.

We must halt the destruction of our forests. Only a few natural forests remain of the natural Earth. Not only should these areas be protected from any human activity, they should be rigorously studied to determine how these systems function so they can be recreated. The remaining forests that have not yet been raped by our corporate industrial complex must remain intact. Not only to protect the wildlife that depend on these holdovers from the natural world, but to preserve these complex ecosystems for the rich genetic diversity and the wide variety of medicinal herbs and plants that have yet to be identified.

Cease all Mining of the Planet

Mining is destructive and a very heavy polluter in every aspect. All over the world corporations are digging up minerals that belong in the Earth while polluting nearby water resources and ecosystems.

We can do quite well without gouging the planet for metals and minerals. All the raw materials our society needs have already been unearthed. We should recycle all the metals and products that are already here. This will happen automatically when we use government regulations and resource-degradation tax structures to make corporate values compatible with environmental and human labor responsibility. This will spark civically responsible corporate-funded research to quickly uncover new methods to extract the highest efficiency possible from new product and manufacturing design. This would result in raw materials being extracted from products that are recycled. This can be achieved by including incentives we have already covered (taxing civic costs and biosphere degradation rather than labor while rewarding new technologies with incentives).

Empty all Landfills and Recycle Everything

We callously toss out any type of trash without giving any thought to where it is going. We need to create a whole new industry that will identify and excavate the world's landfills and safely neutralize, decontaminate, and recycle everything that is found within them. The technology exists to do this, and it is called Thermal Depolymerization Process, or TDP.

TDP can handle almost any waste product imaginable, including turkey offal, tires, plastic bottles, harbor-dredged muck, old computers, municipal garbage, cornstalks, paper-pulp effluent, infectious medical waste, oil-refinery residues, biological weapons such as anthrax spores, sewage and more. TDP converts and recycles this wide range of waste materials into usable oil, gas, carbon solids, mineral solids, metal solids, and distilled water. Other byproducts of this process could also include fatty acids for soap, tires, paints, and lubricants as well as hydrochloric acid that is used to make cleaners and solvents.[325]

All Garbage to be Sorted and Recycled by all Residents, Governments, and Businesses of the World

Businesses, restaurants, hotels, apartment complexes, and homes throw out trash destined for landfills where it will remain for thousands of years. Non-degradable debris, household refuse, industrial wastes, organic compounds, electronic equipment, and other trash (all of which is 100 percent recyclable) is mounting, leaking toxins into groundwater, and filling up landfills. When old sites are full, new ones are opened. Incinerators to burn this overflow of consumer packaging and toxic wastes send plumes of soot into the air. (These deadly plants are usually built near low-income neighborhoods that do not have the clout to stop them.)

How long can we produce mountains of trash? When will enough be enough? Everything can be recycled. In the long run it will be more economical to run an effective program that sorts and processes all leftover trash to be reused by our manufacturers. Businesses that perform tasks, such as de-manufacturing for example, should be the types of businesses that receive subsidies and tax incentives. (The service economy that we will examine soon will promote this type of work.)

Create Incentives for Sustainable Power Sources

Nuclear power only supplies about 20 percent of U.S. electricity. What this means is that simply by conserving 20 percent of the power we use, this entire industry could be eliminated. With more efficient use of electricity and conservation we can dramatically cut our demand for electric power.

But it would be even more promising if the government could support alternative power industries, especially hydrogen. Phasing in alternative sources of energy such as hydrogen, biomass, wind, and solar can improve all dependency on centralized power distribution. Tax credits and government subsidies for these life saving technologies would encourage their development.

Marvin Harris wrote, "Only by decentralizing our basic mode of energy production—by breaking the cartels that monopolize the present system of energy production and by creating new decentralized forms of energy technology—can we restore the ecological and cultural configuration that led to the emergence of political democracy in Europe."[326] The key to true freedom lies in self-sufficiency, not only with food production and water, but in power. The more dependent on centralized power citizens are, the more control they give away to the system and the more exploitable they become.

It takes energy to create these new technologies, and our current primary fuel source is oil, which will be gone in less than fifty years. Now is the time to begin creating the infrastructure and technologies for the next generation in power production.

Shut Down all Nuclear Power Plants

Worldwide there are currently 438 nuclear power plants generating 16 percent of the world power. In a situation of militaristic or domestic crisis, nuclear power facilities would be extremely vulnerable and dangerous. Plus, one or another of them is destined to malfunction someday—it is just a matter of time. The devastation this would impose on all life is unspeakable and unforgivable. More of our world's energy output is wasted than is generated by this dangerous technology. Furthermore, it is creating an incredible amount of lethal spent fuel that is impossible to store safely.

We must invest in alternative sustainable sources of power and find ways to conserve our massive consumption of energy. Nuclear power plants have proven themselves unworthy, and must be safely terminated now, while we have the time and the resources to do it.

Neutralize all Nuclear Wastes

Nuclear waste stockpiles are increasing at the sites in which they are created. The entities that control these power plants are seeking to store the wastes they create at other locations. Here the wastes would remain dangerous for thousands of years.

If the technology does not presently exist, we must employ all of our human technical expertise to solve the problem of spent nuclear fuel by rendering them benign. To protect future life we must invent a method that will safely decontaminate the radioactive byproducts of nuclear fission.

Neutralize all Biological, Chemical, and Nuclear Weaponry

The same holds true for huge arsenals of dangerous chemicals and toxic compounds the militaries of the world are stockholding. Many of these dangerous arms are beginning to show signs of disintegration. The military would like to incinerate them, which would send dangerous gasses into the atmosphere.

We must devise methods to render them benign for the future of all life on Earth. There must be some method we have yet to discover that will break these substances down into their simple, harmless components.

Eliminate all Space Programs

It is truly a horror to tally the amount of resources we violently hurl into the ether of space in which our planet floats. We are culturally programmed to pioneer, explore, conquer, and colonize. However, the myth that human beings could permanently settle in some place other than Earth, of which we evolved to be and stay a symbiotic life component, is beyond absurd.

There are far better ways to use our resources and human ingenuity than exploring space. Another unnerving aspect of sending craft to space is some of them are loaded with nuclear materials. We know that accidents will always be an unfortunate feature of this industry. One nuclear mishap in the form of a space launch would have dreadfully devastating consequences for all life on Earth.

Cease Intensive Agribusiness Practices and Liquidate all Land Monopolies

We have allowed small groups of individuals to monopolize the source of our very sustenance. Today, merely 2 percent of the population produces the food we eat, and an even smaller percentage owns it. These agribusinesses have betrayed our trust and are destroying the very basis of our own personal nourishment as they poison our soils and push the farmlands to their limits. Large corporate control over the most important factor in life and vitality is making us sick.

We must return the soils of our world to those who will protect it. The local farmer respects his land too much to allow it to become poisoned and devoid of life. The foundation of every human society that ever was, including ours, is the land we live upon. The people and nature deserve this space, not a few select members of the wealthy caste.

Localize Food Supply

On average, a bite of food now comes from fourteen thousand miles away. You can now eat a meal that was supplied from every corner of the globe. But by allowing corporations to export/import, we must confiscate land somewhere that could be growing food for the local population, as well as provide more jobs and a stable way of life.

Sending food around the world creates unsustainable urban areas that are very land intensive with regards to the amount of foreign land they require to feed its populace. Most people don't really care where their food is coming from, they will just buy it if they see it and they want it. Corporate morality (an oxymoron because corporate form is innately incapable of feeling) allows agribusiness to reinvest this apathetic cash flow to procure more and more land, no matter who is living on it or what local person is currently depending on the food grown in that field. Many times, someone in another land will starve so we can have a wide variety of food and eat more of it than we should.

Allowing an anonymous source to provide our sustenance separates us from the connection with the land upon which the food was grown. This leads to a lack of accountability. Food and the quality of it are the source of our well-being. It is our immunity. It is our strength. We will weaken that bond if we allow the quality of our nourishment to become compromised to help someone else's bottom line, and nowadays that is usually a mega-corporation. This power is becoming concentrated in a smaller number of hands. These powerful entities are not as concerned about nourishing their fellow human beings as they are their bank accounts.

We must move away from a globalized food market. In just a few years it will be increasingly difficult and expensive to import/export food. If we taxed long hauls of food items and rewarded any food grown for the local market, we would help localize the food supply and conserve resources and energy. This would help populate only those areas that can realistically support themselves. Reducing our dependency of foreign agriculture, trade, and petroleum reserves will largely hinge on our ability to feed local populations with local food.

Educate all of Humankind on the Devastation of the Environment by Animal Industries and Manufacturing Facilities

Most of the human population is totally oblivious to the effect that meat production has on our planet. Most people believe that cows and pigs have the same impact on the environment as wild animals do.

The fact is that if we could eat less meat as a culture, all those fields that are currently being raped for animal feed, season after season, could be returned to forested land. Streams and lakes could once again run clean. The oceans could replenish. Use of energy resources such as oil, coal, electricity, and gas would drop dramatically. Land values would drop, bringing back sustainable agriculture, and along with that a more peaceful world. This will not happen unless our society makes this critical information abundantly available to the global population.

Restore Life to all Waterways and Reforest all Available Land as Public and Global Policy

Our culture views rivers and forests as facets of the Earth that are outside of their reality. To some urban residents, it may almost seem as if they don't even exist, except perhaps as places to utilize for recreation, as if it were a product.

Healthy forests provide flood control, climate stabilization, water purification, and wildlife habitat, as well as replenish the topsoils we may now be destroying. What price could be put on those services? We must *invest* our resources in returning the world to its natural condition. Life must be allowed to occur without interference. We need to develop specialized teams that can recreate the original natural conditions for life to evolve in every available sub-ecosystem that can be found, and to monitor each system for integrity and genetic (diverse) strength.

As the human race reduces its population, it must be a new global policy that we return depopulated urban areas to nature as best as possible.

Seven Principles for Sustainable, Healthy, and Clean Water Resources

We fight water pollution with home filtration units and bottled water (that add vast amounts of plastic to landfills) as sewage plants chemically treat municipal wastes before they dump it into rivers for the next community to drink. We must begin now to stop the severe degradation of our water resources. These are the Seven Principles:

1. **Water is a basic human right**—Nothing is more essential to sustain life than water, other than air. And not only is it a human right, it is the natural right of all species.

2. **Water belongs in the public domain**—As the universal and global right that it is, water resources are a collective responsibility and belong in the public trust, protected at every level of government of every nation. The citizens are the best advocates because they live in the communities where they depend on clean water for their lives.

3. **Water is not a commodity**—Corporation around the world are seeking to control the distribution and supply for profit, and this is antithetical to the basic right to water. Water should be heavily regulated by laws strictly

enforced. All corporations must be banned from including water resources in any trade deal.

4. **Water resources must be protected at all costs**—We must conserve our water supplies and reclaim those sources that have become polluted. The simple single celled algae can be utilized to treat sewage of all kinds. Algae can metabolize and render benign wastes from animal industries as well as human wastes. Small-scale treatment facilities could process, treat and recycle local wastes for use on gardens and farmlands. Wetlands and riverbanks also offer a very inexpensive and effective method for filtering water. They not only remove bacteria that can be harmful, but they also deal very effectively with viruses, microbes, and chemicals. This can be done by drawing water out of wells alongside rivers rather than directly out of the river itself. This has already been successfully demonstrated in Europe.[327] We must separate organic wastes from chemical wastes. We all need to become educated on these issues and must be taught how to implement these solutions in easy step-by-step methods. Wetlands and rivers must be restored and natural water systems must not be tampered with or altered from their natural state. Watersheds cross national borders and so should be treated as whole ecosystems by all governmental jurisdictions and authorities.

5. **Dam constructions must be halted as all large dams are deconstructed**—We have learned how dams destroy ecosystems, species, and cultures, as they pollute and snuff out natural systems. To renew these systems and save those about to be destroyed we must declare a global moratorium on all dams as we begin the work of reversing the damage that has already occurred.

6. **Establish local Water Governance Councils**—These councils will be trusted with the local preservation and conservation of the public source of water. They would have the power to supersede all preexisting trade agreements and property rights for the common good of the local communities who will retain all rights to the local water resources.

7. **Establish a Global Water Convention**—This will serve to create and establish international treaties and agreements to gain international cooperation for the provision of clean water for every human being.

Harvest and Use Algae, Kenaf, and Hemp Instead of Trees

We are currently tearing apart living ecosystems to communicate with each other. The conventional source for lumber and paper products is white pine. Alternatives such as kenaf are practically ignored by those industries that earn obscene sums of cash from ripping apart forests. Hemp is promoted as the "evil weed" by vested interests whose polluting and heavily subsidized low-employment industries would be devastated by its popular use.

There are other sources that can produce more paper using less land. Algae ponds could be harvested for fiber that could be used to manufacture paper. Each acre could produce up to 134 tons of fiber per year[328] (as compared to the best tropical trees producing a one-time harvest of only 16 tons while rendering the land useless for future use). One acre of algae production can also produce 3,400 gallons of oil per year. Algae can also be dried, compacted, and used as biomass fuels.

Kenaf is another plant that shows remarkable promise. Kenaf only takes about 150 days to grow a crop, while White Pine, the traditional tree, takes about 14 to 17 years. Kenaf is oil absorbent and so it can be used to clean oil spills. Kenaf offers a high yield of 5 to 10 tons of dry fiber per acre, which is three to five times that of pine. Kenaf can be rotated with other crops to help discourage pests and the subsequent use of pesticides. Kenaf requires 15 to 25 percent less energy than most other crops. Existing mills can be easily converted from processing lumber to processing kenaf instead.

Hemp is another excellent source of the highest quality fiber. One acre of hemp can produce as much fiber as 4.1 acres of trees. Hemp was used for centuries to manufacture rope and twine. Nylon is presently used, but the petrochemical source is polluting and once again produces few jobs, concentrating the wealth into the hands of a few. Hemp can replace cotton for making clothes and rugs and other textiles. Not only is hemp softer, warmer, and more absorbent than cotton, it requires no chemicals. (Fifty percent of the agricultural pollution comes from cotton.) Burning petrochemicals releases ancient excess stores of carbon dioxide. Burning newly grown biomass (e.g. hemp and algae) releases only the carbon dioxide consumed by the plant during its life. Building materials can be constructed with hemp fiber. Engineered lumber and oriented strand board are stronger than wood. Plastic can be made from hemp seeds. Insulation can be manufactured out of hemp instead of the hideously dangerous materials currently in use. The roots of hemp are sturdy and help to prevent soil erosion and mudslides. Hemp will grow in all climatic zones. There simply is no need to destroy another forest. Indeed, there never was.

Restore Air Quality with Algae Ponds

It seems that there is nothing we can do to reverse the buildup of carbon dioxide in the atmosphere. Most of us seem willing to just accept the fate we have brought upon ourselves and view the degradation of the air we breathe as the price we must pay for Progress.

Large algae ponds located throughout the planet's surface could reverse the buildup of carbon dioxide in the atmosphere and oxygenate our air. One 2.5-acre algae pond can pull 6.3 tons of carbon out of the air and return 16.8 tons of oxygen a year.[329] This is three times the rate of the average acre of forestland. Why is this so? Three billion years ago there was no oxygen in the atmosphere. It was algae, not forests, that created oxygen as it broke down the carbon dioxide molecule, storing the carbon in its cells and releasing the oxygen into the air. Ninety percent of the oxygen we breathe is created by algae. The technology exists to construct these ponds today.

Increase and Modernize Public Transportation Systems

Freeways and large gas guzzling sport utility vehicles are proliferating all over the world. We are becoming so dependent on foreign oil we might as well just invade these oil-rich nations and get it over with. [Author's note: I wrote those words in 1999 before the Bush Administration occupied Iraq in 2003.] As our population increases, the dependency on whatever transportation system is in place becomes greater. In our reality that is cars. Lots and lots of cars.

With our large population, we need to have an effective transportation system that will be able to deliver goods and services without polluting our environment. We will need to be able to move around cheaply and easily while we decrease our population base. Rail and bicycles offer some hope, but new technologies need to be developed.

RELATIONSHIPS

Eliminate Capitol Punishment

The United States is the only western industrialized nation that has the death penalty. When the state kills someone, more than just the death of one (possibly innocent) person occurs. We send a message to future generations that murder is an acceptable method of retribution. We are saying that the eye-for-an-eye biblical philosophy is valid in our modern reality. And when our society validates such unsavory and barbaric means of punishment, we will find that the next generation will adopt those means and the mindset that goes along with it. They learn not that they should be law-abiding citizens lest they be executed; they learn that execution itself is an acceptable way to deal with others. If an individual is deemed to be dangerous to society, they should be incarcerated for life.

Furthermore, it has been determined that one out of seven inmates on death row are innocent.

Study all Violent Offenders to Determine the Underlying Cause of their Malevolent Behavior

Prisoners are people who cost millions to society for doing nothing productive while serving their sentences, except learning about criminal techniques from other inmates.

People who commit violent crimes are some of the most valuable resources we have in preventing crime from happening in the first place. What prompted this individual to commit this atrocity? Would they (or others like them) do it again? What social factors encouraged this individual to become violent? Was it an inherited tendency? Why, why, why? Criminals should be interviewed and their information documented to help us eliminate the conditions prevalent in this person's life that led to the manifestation of their violent behavior. We must always remember that criminals are a symptom of our societal ills, not the cause!

Decriminalize all Drugs and Set Up Drug Counseling and Education Centers

Drugs are illegal, except for certain drugs that are manufactured by powerful corporations that have political clout, such as liquor, tobacco, and caffeine. Therefore, possessing or selling a small amount of hemp for smoking could potentially land you in prison. Addicts are not typically violent people, but we will lock them up in our prisons alongside violent people, where they will learn violence. As it was with prohibition in the 1920s, so it is with the prohibition of other substances today that alter conscious states. Since networks will develop to provide the products that are in demand, the very fact that drugs remain illegal is what boosts the profit margin and creates criminals.

By decriminalizing drugs, we will quash the power base of organized crime. We need to invest our energies into more productive methods such as education and addiction-counseling centers. We need to help people who are addicted to drugs, not treat them like they're inhuman or expendable. Decriminalization will also help restore a peaceful way of life for innocent people who live in such parts of the world as Columbia. The innocent usually pay the highest price when caught in the crossfire of the War on Drugs, paid for by the American Taxpayer.

Teach Environmental Restoration Skills and Sustainable Farming Skills to all Prison Inmates

Why do they call prisons "correctional institutions" when they are cells where people waste their lives away, totally isolated from society? This is not a place to learn social skills or how to peacefully interface with our society. Prisoners are not taught any valuable skills at all. A prison is a criminal networking and training center. They watch violent television shows and learn about how other criminals operate. Inmates are mostly discouraged from reading and are coerced to live in what one prisoner has termed the "TV universe." Inmates are also exploited by corporations as cheap slave labor.

The most valuable professions of the future Earth will be those that enhance our beautiful planet. Reforestation, Organic Farming, Toxin Sanitation, Wildlife Enhancement, Nuclear Neutralization, Chemical Processing, and Waste Management will not only be lucrative opportunities, they will be some of the most important work anyone could perform. This is what all inmates should be required to learn, and where their focus should be, in addition to regular counseling and study.

Shelters for the Homeless, Abused, and Elderly

Millions of oppressed individuals live here in the U.S. and throughout the world without shelter and little food. How can any nation be proud of itself when its own fellow citizens are abused and their grandparents are destitute?

Ultimately, are we to be judged on the basis of our economic success, or with how our culture treats those of us who are in need? We should give the downtrodden among us at least the same creature comforts we give a dog (basic nutrition and shelter). In these shelters, people could receive drug counseling and be studied to ascertain why they are homeless so we, as a society, can prevent those conditions which caused these individuals to be unable to survive in our society in the first place. Then, we should train those who are trainable to function as valuable members of our society, much like my proposal to mandate the education of prison inmates.

Free Confidential Birth Control and Education for all People

More people today will create more suffering tomorrow because our ability to feed the billions that are here today is insufficient, and it will not improve. And yet fundamentalist religious orders, including the Vatican (which has tremendous influence on such matters) are always attempting to limit access to birth control and education for women. Furthermore, we currently give proportional tax breaks for people who have children.

In this modern era, reproductive freedom should no longer be considered a "God-given right." We need to encourage the use of all forms of birth control and give women the right they deserve over their own bodies. Humanity and all life on the planet will benefit immensely if all humans have complete control over their reproductivity. By limiting and decreasing our population, we increase the quality of life for future generations and the survival of our species. Birth control and education is criticized by certain groups and individuals as unnatural or immoral. But our exceedingly exorbitant numbers are due to unnatural means (agriculture) and may require unnatural means to balance its effect. We need to encourage the global population to have small families, or preferably no children at all. We must do this to protect future generations from violent and certain mass suffering and obliteration in the fight for survival and dwindling resources.

Free Education for all to Nourish Artistic Expression, Natural Talents, and Cultural Diversity

All around the world educational systems teach children to become specialists in the western Dominator urban environment. Our current educational system is primarily constructed to suit the needs of corporate employers who seek conformity, predictability, and mono-focused human units to plug into receptacles where they will predictably perform and produce within the framework of the corporate machine. Educational institutions should exist to serve the individuals who are receiving the education, not to make them easier to exploit during their productive years.

An educated citizenry can only serve to benefit society, and so education for all must be free. Educational systems need to be decentralized and diversified to fit local cultures and environments so they are sensitive to not only local culture, but also the local environment to encourage local sustainability and independence from foreign and corporate dominance.

Society as a whole needs to embrace the fact that different human beings possess a wide variety of talents in various degrees in other areas than the "3 Rs," although these areas are important too. In fact, there are at least seven categories of human intelligence, although the vast majority of our school's course work emphasizes proficiency in primarily two. What this means is that people who have much more to offer others and themselves in the way of life-enhancing skills are forced to conform and perform in fields of study in which they may have no special talents or interests. These are the seven areas of human intelligence:[330]

1. **Verbal/Linguistic**—These talents are found among our writers and poets.

2. **Logical/Mathematical**—These talents are shared by our chemists and scientists.

3. **Musical**—Abilities such as these are critical for cultural expression.

4. **Spatial**—People with these innate skills have a variety of applications such as navigation, topology, sculptors, artists, architects, chess and physical sciences. The ability to utilize imagery to envisage solutions to complex problems is also a valuable skill for many human endeavors.

5. **Bodily/Kinesthetic**—This is a critical intelligence for sports, dancing, musical performance, arts, construction, physical health, and acting.

6. **Intrapersonal**—This category of intelligence involves the ability "to detect and to symbolize complex and highly differentiate sets of feeling." These people can tap inner experiences and feelings to express their inner selves, or may help others do this by example.

7. **Interpersonal**—People who possess a high level of this type of intelligence can read the intentions and desires of other people, even when the others attempt to conceal their true intentions. High interpersonal intelligence would also allow one to influence the behavior of others. These people make good leaders.

Two other areas of human intelligence might be **Mystical** and **Pattern Recognition**. Perhaps there are others. All people possess these inborn intelligences, to varying degrees.

There are many functions a person can fill in society or an expression of culture that require great precision in more than just one mode of intelligence. Educational systems should focus on the individual's interests and inborn talents while cultivating and balancing all forms of innate human intelligence for the purpose of creating self-fulfilled, inspired, resourceful, and productive people.

Create Regional Cultural Events

We watch a lot of television in our culture. Our creative minds are stifled in our culture by commodification, which encourages consumption and requires passive non-participation.

Culture and artistic expression are part of the very essence of being human. They create identity and help us define our personal relationship to the perceived cosmos and each other. Music, stories, art, poetry, and other creative forms of self-expression delight us, entertain our senses and provide joy and communion with other people. It is that shared experience that will strengthen our bonds to our past and coalesce our collective vision for the future. Local communities need to cherish those individuals who possess the capability to enhance and enrich the social and cultural fabric of their societies.

We need composers, writers, actors, and other creative individuals and organizations. These creative groups and individuals should be free to create for the benefit of each culture. They should be free of corporate consumerist constraints and bureaucratic governmental regulation, and other forces that attempt to define, limit, and narrow the boundaries of expression for the purpose of mass marketing various products and ideas. Society should cultivate and encourage their talents for the enhancement of the quality of life for all members of a community.

For culture to survive (all human society depends on it), every region and subregion of human community around the world should be allowed and encouraged to create their own identity and nurture spontaneous and diverse human creativity in art, music, and lifestyles within their own society. This is human cultural evolution. It is not a product. It could never be manifested in an instant and set in stone. It is a process and a direction.

Protect the Natural Evolution of Indigenous Cultures

Our culture views indigenous peoples as being in need of our technological and spiritual guidance to help them live a better life. That is the excuse our institutions offer to explain why we destroy the cultures and the way of life of those whose economies and resources are not integrated with ours, allowing for corporate and political exploitation.

Ninety-nine percent of human history was tribal. Indigenous cultures possess the agricultural techniques for sustainable living. We of the Domination culture desperately need to learn and fully comprehend the psychology of tribal structure. They must be allowed to evolve free of any interaction with the current global culture. Only in this way can we preserve the knowledge that they possess about being a natural human. They represent our last link with who we are and where we came from.

Create Incentives for Communal Groups to Become Self-Sufficient

The survival of most communities now depends on outside sources for many goods and services. Reducing dependency on extensive trade will allow local groups to self-determine their own course in culture and law. By mandating tax incentives to encourage local autonomy, we can gradually reduce the burden on national and international financial assistance as well as influence. Communal localized communities should be free to create their own codes and have a vastly reduced tax liability (if not entirely eradicated) if that community becomes truly self-sufficient.

This is one huge step, and perhaps one of the most far-fetched in the mind of a conventional thinker. But this is also one of the most critical and delicate advances in the Cultural Evolution I am proposing. Each of these Permanent Autonomous Zones (PAZ) would be completely responsible for its food, government, culture, transportation, and well-being. Each PAZ should be treated by the rest of humanity as the sacred phenomenon it is—a spiritual sacrosanct experiment in humanity whose evolutionary course and existence would be protected at all costs as they are the cultural seeds for future human evolution.

Furthermore, those zones which have successfully severed their ties and wish to become culturally independent should be granted the right to non-interference by the rest of the global community, provided they never pose a threat to the peace and stability of the rest of civilization. I call this the **Principle of Cultural Non-interference**.

COMMERCE AND GOVERNMENT

Cease all Corporate Donations to Elections

Corporations run the electoral process in most nations. By infusing large amounts of capital, they can influence the policies of the political parties as well as public opinion. If the parties that select the governmental servants are beholden to private financial interests, the politicians we elect will create laws that favor those interests.

The political system must serve to express the desires of the community at large. The authorities must be beholden only to the best interests of human society and welfare. Political forums must be created where all ideas are given equal and fair consideration, with no capital invested from any source towards the propagation of one idea or another. Equal exposure to the widest range of all societal, cultural, and governmental opinions and viewpoints must be accessible to all citizens.

Disallow any Corporate Influence in the Law Making Process

Corporations determine the policies of government agencies by placing corporate-friendly officials into positions of authority in government. Donations to government officials or political parties and the dispensation of gifts will usually bring about laws and regulations that will favor the corporations.

Being a proxy for special interests, especially for the procurement of power or financial gain, either personal or political, should be considered the gravest of societal crimes. When our political servants gather in any size, representation, or capacity, the decisions they are asked to make should be free of any outside special interests other than specifically those interests which benefit only the human societal entity for which they have been selected to serve.

End all Corporate Subsidies

Corporations are heavily subsidized by special tax breaks and government funding of projects that only benefit certain corporate sectors of modern society. This has come to be known as "corporate welfare."

No government authority should be dispensing special favors to any unworthy organization. Subsidies and tax breaks are to be generously provided for those services, products, and technologies that enhance the quality and diversity of life as well as human culture. Those corporations that destroy life or arrest the natural progression of culture for profit should be accordingly taxed or penalized in accordance with the severity in which they degrade.

Restrict Corporations from Broadcasting Rights and Make Television Responsible to Humanity

Corporations currently utilize public airways solely for the dissemination of its own self-serving agenda. Corporations are inherently incapable of fairly and accurately dispensing all sides of an issue, so long as the corporate shareholders have a vested interest in one of the sides. Television's view of reality is warped because it only presents the corporate viewpoint, that humans are only consumers.

Along with the right to broadcast comes the responsibility for equal representation of all views for any social/cultural/political equation. Public access to this technology should be in the hands of a public commission selected by society through a public forum. The forum should be responsible to only the best interests of human society and to see that all factions have a fair voice in the exchange of ideas and facts to assist the public in making unbiased decisions. This rule should also apply to radio, newspapers, and all the mass media.

What humans view and experience shapes their perception of reality. Humanity needs to recreate the medium by eliminating the corporate control over the content and give control of the airwaves for all media to the people. The airwaves are in the public domain and no government has the right to sell them to corporate interests. The content of all publicly transmitted information must be only in the pure interest of the public and not for commercial use.

Reduce Global Trade and Eliminate "Free Trade" Agreements

Today you can buy a product whose parts were manufactured in several different countries. This creates a dependency on other regions as communities lose the ability to produce various commodities for themselves. It also creates an information gap. We are not aware when the manufacturing process of a product we buy pollutes the environment of a different region. We are also not aware when a product we buy was created by slave laborers working under dire conditions. There is no accountability for what we purchase anymore. And as money exchanges hands, it is the larger corporations that make the most profit.

Global trade of commodities that could be produced locally is a total waste of energy through the transportation of goods from one locality to another.

The only way we can create autonomous regions of self-sufficient human societies and communities is to reduce dependence on imports in all sectors of the planet. All regions should place high taxes on trade to discourage foreign manipulation and exploitation. Free trade institutions such as NAFTA and the WTO must be dismantled to return democracy to the world citizens and stop global corporate bureaucracies from exploiting the labor and natural resources around the globe.

De-Globalize all Corporate Structures and Regulate Them Locally

Global corporations will never, by their inherent nature, function for the best interests of any community or human society, for these are the very groups they are designed to exploit. Today, a franchise can destroy the local economy of any community by pricing the local merchants out of business and gaining a monopoly of local retail sales. They can underpay their employees and push out locally owned businesses.

Every community should have control over the business transactions that will have a local impact to ensure that those transactions are in its best interests. All corporations should be subject to publicly debated and sanctioned charters that need to be renewed and reviewed regularly. Localized regions must disallow imports, or heavily tax imports of services and products that can be locally performed and produced (especially if those goods and services can be purchased with local currency). To help with deglobalization, corporations should be subject to the following four preconditions:

1. Restriction from owning stock in other corporations;

2. Restricted from dissolution until all disputes have been settled;

3. Stockholders should be held responsible for all the liabilities of the corporate stock they own;

4. The local impact of the behavior of every corporation that does business in a community should be reviewed regularly to renew local charters that permit that corporation to conduct business in that locale.

Eliminate the Corporate Assumption of the Status as Constitutionally Protected Human Beings[331]

As we have thoroughly covered in Section III, *Aspects of Domination*, a corporation is not a human being but an agreement between human beings; therefore, it is not deserving of the rights that a real human being has. But in our reality, a corporation is considered as having the same rights as a flesh and blood human being. If a corporation commits a criminal act that results in harming somebody, the owners of that corporation are not liable since the actions of the corporation were not performed by the owners. They cannot be held responsible for injury, death, pollution, destruction, or any other criminal acts that would put a real human into prison. The owners of a corporation use this concept to maximize profits without responsibility to the community. The owners of every corporation should be responsible for what their corporation perpetrates on other beings in their name. As we know, the situation as it sits now is completely the opposite. Corporations enjoy the rights of humans without the responsibilities.

The legal status of a corporation should be "artificial entity" as it was until 1886. This would have the effect of making businesses more responsible when offering services, creating products, or employing humans to function within its framework.

Create a Global Reference Currency[332]

Right now the dollar is backed by and based upon nothing. The dollar was once a note that represented (in theory) a dollar's worth of gold (until 1971 when President Nixon removed the gold standard). All currencies of the world are tied to and measured by the U.S. dollar. But currencies should be specific to the arena in which it is traded. The dollar should really have no value outside of the sphere for which was created—within the borders of the United States of America.

A global currency for global trade would stabilize the economies of the world and prevent money markets and speculation. In his book *The Future of Money*, Bernard Lietaer writes that a "Global Reference Currency (GRC)" is a "currency which is not tied to any particular nation state, and whose main purpose is to provide a stable and reliable reference currency for international contracts and trade. Furthermore, I will propose as unit of account for one particular type of GRC the 'Terra,' which aims at firmly anchoring that currency to the material/ physical world . . . a standard basket of commodities and services."

The global unit could be called a Terra, the value of which could be agreed upon by the nations who might trade with that currency. For example, one Terra could equal a "basket" of an international/global set of measured commodities. This negotiated set of goods (oil, wheat, copper, gold, etc.) would be inflation proof. A Terra, would be, in a sense, a warehouse receipt. It would carry a demurrage charge, or sustainability fee, for the storage of the goods it represents. This would have the effect of discouraging hording of the currency itself. Cash would flow and keep commerce healthy. When the need would arise to change Terras into a national currency, the value of one Terra would be whatever the value of the "basket" these pre-determined measured products (and services, too) would be in whatever currency you are comparing it to. This would be the epitome of stability. So if the commodities represented in one Terra would cost $457 to purchase, that would be what one Terra would be worth to an American trader.

If trade is to be done on a global basis, it should be done with a stable global currency and the Terra would be a currency for primarily international trade. While the vision advocated in this book would discourage trade on a global level, it cannot be stopped tomorrow without huge disruptions and global calamity. The Terra would be a temporary device for global trade until it would become unnecessary for future generations to waste precious energy by shipping products from other regions. National currencies would still remain in use for trade within the borders of each nation.

This concept is endorsed by Nobel Prize winners and economists the world over. In summary, the Terra "would enable a private initiative to address five key international problems that businesses are experiencing today:

1. It makes available to businesses an international standard of value.

2. It reduces the cost of completing some counter trade transactions.

3. It provides an insurance against uncertainties deriving from international currency instability.

4. It systemically reduces the possibility or seriousness of a global recession.

5. It structurally resolves the conflict between long-term ecological sustainability and financial priorities built in by the conventional currency system."[333]

Localize Money Flow by Using a Community Currency

Just as using an international commodities-based currency for international trade is prudent, and using a national currency for trade within a nation's borders is logical, so is it wise to use a community currency for trade within that community.

If you use the same monetary tool to trade oil from the other side of the world as for paying your babysitter, this is going to cause poverty and scarcity within the majority of local communities. International money markets encourage global investment and speculation, and overlook the value of the relationships that exist in our communities. International currencies are the antithesis of localized sustainable societies. It encourages outside vested interests to own local businesses and then base their business decisions upon extracting the greatest profit from that community and not what is best for the community. They have no stake in the community that provides the labor force and for which it is commissioned to serve.

Autonomous regions and local communities can reduce outside dependence by localizing trade and taxing imports, fostering local opportunity, and discouraging outside corporate control of local businesses. Issuing currency that is redeemable only within a region or community will keep the value of work and prosperity circulating within that community. This creates work and local wealth and could replace welfare. As we know, "welfare is a compulsory transfer of resources from the rich to the poor via taxes,"[334] and a local currency would be capable of providing a method of compensating people for work they do in their community, allowing them the right to a livelihood. Local businesses could easily accept the currency and would have the added benefit of helping them compete with the large chain distribution mega-stores.

Currencies such as this could not replace national currencies, but enhance them, and as such would be called "complementary currencies." They would exist to perform social functions that national currencies cannot perform.[335] This would also serve to create new wealth and new services that would serve the interests of the community, not some distant money vacuum. Environmental cleanup, cultural and artistic expression, public transportation, elderly care, youth mentoring, child care, and housing rehab are just a few of the possibilities that local money systems could solve.[336]

Most of these currencies are mutual credit. This system creates money by the participants through the rendering of service debited by time. When you perform a service for a fellow citizen, your account is credited one hour, and when a service is done for you your account is debited by the same amount. There are several variations of this that would include:

- varied exchange rates for services that may be worth more than others;
- one unit of currency could be tied to another currency;
- and it could use paper bills or a central computer.

There are many thousands of successful systems like this operating worldwide today, as well as since recorded time.

This would provide the balance to the commercial economy by creating a community economy. Bernard Lietaer in his outstanding book, *The Future of Money*, refers to these as the Yang and the Yin of currencies, respectively. The Yang economy is the financial capital we are now familiar with that we use for commercial transactions and is based upon competition and scarcity. The Yin economy would provide the social capital for community transactions and would be based upon cooperation, self-sufficiency, and the establishment of relationships, something that is healthy in any community or tribe.[337]

Halt all Sales and Gifts and Cease Production of all Arms, Munitions, Weaponry, and Devices of Destruction

The U.S. government supports arms sales by U.S. corporations to any buyers (including rogue nations that may turn those very weapons against our own troops) as a means to support our economy and provide jobs. The American taxpayer foots the bill to pay the corporate merchants of death, killing millions of innocent men, women, and children all over the world.

Small foreign dictatorships do not need these weapons and we should not be providing them, no matter how good it is for our economy. Peace is never served by violence.

Change Election Procedures to Instant Runoff Voting and Offer Incentives to Vote

Our current system allows a minority of the people to impose its values, belief systems, and priorities on the majority of the citizens. It's no secret that corporate interests have a big stake in the control of the major parties. Under the current winner-take-all system found in the U.S., if there are several choices at one end of the political spectrum, the opposing spectrum gains an advantage. This is because the votes on more populous and diverse end of the spectrum are spread thin. The majority may vote, ideologically speaking, in one direction and not even count because there was less competition amongst the candidates on the minority end. Why should one end of the political/social/cultural/economic spectrum of thought be penalized on the grounds that they offer more diversity in ideas? The wealthy class retains control over the process and exerts control over not one, but both major parties.

To accurately select those officials to serve and represent human society, we need to redesign the method in which we choose those individuals. Most western nations have some sort Instant Runoff Voting (IRV) built into their election procedures. In a system such as this, all voters would have the option of selecting a second choice just in case their first choice is not selected. This would allow all voters to vote how they truly feel without fear of accidentally giving a drastically opposed moral and ethical perspective the advantage to prevail just because one faction of ideological thought offers a greater variety of choices. In this way, the true will of the people will not be subverted in situations where there are too many similarly positioned candidates to choose from.

Voting days should be national holidays to encourage working people to vote. Registering to vote should be easy to do, and care must be taken to allow the homeless to vote too. Information about candidates' positions and other items up for vote should be plentiful and abundant. An informed public would be good for democracy.

How we vote could also be up for a vote. Anything can be improved upon, even the basic structure of the system itself. Nothing should be created that cannot be altered or eliminated. Perhaps some day we could consider having citizens vote not for individual human beings, but for ideas and principles. Maybe it would be to our advantage to select platforms and parties to fulfill societal obligations. The selected parties might then choose the individuals who would be responsible to fulfill the mandates of the people. In this way we might avoid the modern pitfalls of postmodern Kings and Queens who occupy higher governmental posts.

Tax only Consumption, Pollution, and Resource Depletion, not Labor or Corporations

In our system we tax the workers. The wealthy can find tax exempt investments and shelters while the common worker gets a big chunk taken out of their check every month to pay for things that will not benefit them at all, and in many cases the money they pay goes for things that put them at a greater disadvantage.

We do not need to tax individuals who just make enough to pay their bills and own or rent a home. The bills and the ills of our world would justly be financed by the ones who are responsible for and profit from these maladies—corporate structures and the well-to-do people who own them. If we shift taxation over to energy consumption, it would make wasteful energy more expensive relative to labor. This would encourage the creation of jobs and reduce our dependence on polluting industries that tend to benefit only a handful of investors.

It may be considered quite unfortunate that the value intrinsic in our culture of other life and other communities of life are of little value unless we appoint one in the form of a number that corresponds to a monetary value. And as the saws rip down entire forests to the bare nub, the value of the harvest is found in the exchange of financial reward for the perpetrators and nothing for what the forest, stream, fish species, or grasslands provided previous to their annihilation. The loss of natural resources is not calculated in the Gross Domestic Product (GDP), and so the loss of the services these resources provide is not counted.

A quick accounting of "services provided" for society would show that much of what we receive for free is not appreciated as much as it might be should we someday find ourselves lacking for it. Thick forests, fertile soils, oceans teeming with life, clean air and the regeneration of oxygen in a toxin-free atmosphere, filtered fresh water, and a strong genetically diverse selection of a wide variety of food plants came at one time at no cost.

Furthermore, we include the costs of our diseases, criminal system, toxic wastes, highway construction, trash collection, and other malevolent symptoms of our decline as growth, in black ink on the positive side of the ledger sheet. Just because money changes hands does not mean the economy is moving in a positive direction. Degenerative Growth (DG) is the profit made by entities that precipitate a loss which is shared by another segment of the global community of life. We must learn how to subtract DG from GDP and levy powerful taxes on all of it. DG is all that our governments should ever need to tax to repair our planet and operate for the common good of all people and all life.

Some examples of DG would be chemicals, fuel, nuclear power, destruction or depletion of ecosystems, and long-distance trade. If these and other DGs were to be identified and taxed, whole industries would not collapse—but they would change their behaviors. People would not cease to purchase consumer goods, but they would alter their choices in products. Manufacturing companies would not grind to a complete halt, but they would adjust their product designs and revise the processes by which they acquire raw materials.

Replace Product Sales Economy with Customer Service Economy

If products were used by the consumer base on a service basis rather than a buy/use/discard basis, we could eliminate the tendency for the polluting and degrading type of linear thinking most commonly found today. Instead of marketing products, businesses would be selling product performance and customer satisfaction. Businesses would retain ownership and lease rather than sell products that are eventually replaced. This would encourage industries to create products that are durable and designed to be deconstructed and reused as raw materials. In this type of economy we would find that products would become the means by which transactions occur, and not the end. It would encourage industries to create products with long and useful lives, and not constantly striving to make the last batch of products as obsolete as possible to encourage the consumer base to restart the linear process of buy/use/discard as quickly as possible.

Invest in Natural Capital

As dwindling resources are beginning to become more difficult for easy exploitation, many businesses are coming to the conclusion that a healthy biosphere, as well as social stability, is a valued commodity in their endeavors to produce financial profits. In other words, a healthy planet will produce resources that will be more abundant, and a content and healthy population will create a more productive work force and a stronger consumer base. Corporations will be rewarded for making contributions to biosphere viability.

Create Incentives for all Sustainable Technologies

As it is, our government subsidizes traditional conservative industries that have powerful lobbies in Washington D.C. This type of behavior occurs all over the world in all types of countries. Mining, lumber, and manufacturing plants all receive considerable benefits from the local governments that in turn provide powerful backing for the local politicians.

Sustainable low environmental impact technologies will be mandatory to a clean environment for future generations. We must encourage this type of research and development to help us humanely cope with the exorbitant numbers of us. We can do this by taxing polluting industries and rewarding environmentally protective oriented businesses. In this way we can encourage and utilize the knowledge and the talent of the best of us to create solutions that will benefit all mankind, not just financially reward the select few.

Create Peace, Cultural, Health, and Environmental Branches of the World's Governments

Corporations are able to dictate legislation that allows them to do whatever they want to create profit. That power must be returned to the citizens of the world.

We need to create and lavishly fund an interconnected network of governmental branches and agencies on a global basis to protect and restore the Earth. These organizations would be responsible for:

1. Education on health, nutrition and environmental issues, particularly what impact various human activities have on our biosphere;

2. Universal health care;

3. Global Peace;

4. Creating permanent autonomous zones and regional/local cultural evolution and independence;

5. Population control;

6. Breaking up agricultural monopolies;

7. Dividing and issuing land to families for organic subsistence farming and regional and community markets;

8. Holding polluters accountable;

9. Limiting international trade;

10. Cleaning waste sites and dumps and recycling all refuse;

11. Regulating corporations and creating a service economy;

12. Eliminating the use of chemicals, toxins, genetic technology, and irradiation of the food supply;

13. Sanitizing and cleaning municipal water supplies and systems;

14. Oxygenating the atmosphere and removing air pollutants;

15. Restoring natural conditions and life to barren waters and lands;

16. Rendering toxins and weaponry benign for safe burial, storage or recycling;

17. Permanently closing down all polluting and nuclear industries;

18. Creating an environmentally benign source of energy;

19. Creating an environmentally benign method of transportation;

20. Switching all paper resources from trees to more environmentally friendly sources such as algae, kenaf, and hemp;

21. Promoting and creating new wealth through the creation of a:

    a. Global Reference Currency based upon a variety of commodities, and

    b. variety of local and community currencies based upon time-sharing and cooperation.

◆ ◆ ◆

The Memes of the Cultural Evolution

These are the primary memes of the Memeplex of the Cultural Evolution. Each piece is a critical component. Most of these memes would have a tough time surviving alone in the Domination culture because the Domination culture is a memeplex with its own memes.

Members of the Cultural Creatives subculture, which now encompasses over 25 percent of the population and is growing, is promoting at least one of these memes individually. One purpose of *Cultural Vision* is to help quicken the public's realization that all our multitude of causes in the name of health care, environmental integrity, human rights, and social justice are based upon one or more of the three primary values. All memes are based on values.

The Domination culture is just one culture. It will have to transform itself to survive. The other purpose of *Culture Vision* is to help quicken the transformation before it collapses. It must evolve now.

To hasten the evolution, all the causes of the Cultural Creatives should be championed by the others. It's not just "Worker's Rights," or "Save the Whales," or "Stop Genetic Modification," or "Health Care for All." This is a cultural values-based evolution that will alter our attitudes towards life on Earth itself, other people(s), and our local communities. Thank you for being a part of the Cultural Evolution.

For easy clarification, here is a quick review of the Memes of the Cultural Evolution:

Health

- Public Education on Nutrition and Exercise
- Guaranteed Health Care for Everyone
- Biodiversify Food Crops
- Halt Chemical Treatment and Genetic Modification of the Food Supply
- Vastly Increase Soil and Food Quality
- Integrate Holistic and Allopathic Modes into Health Care Systems
- Integrate Mind/Body Disciplines into Educational and Health Care Systems

Environment

- Increase Taxes on all Polluting Industry—NO Loopholes
- Vastly Increase Funding for Climatological and Dynamic Oceanographic Research
- Halt all Deforestation
- Cease all Mining of the Planet

- Empty all Landfills and Recycle Everything
- All Garbage to be Sorted and Recycled by all Residents, Governments, and Businesses of the World
- Create Incentives for Sustainable Power Sources
- Shut Down all Nuclear Power Plants
- Neutralize all Nuclear Wastes
- Neutralize all Biological, Chemical, and Nuclear Weaponry
- Eliminate all Space Programs
- Cease Intensive Agribusiness Practices and Liquidate all Land Monopolies
- Localize Food Supply
- Educate all of Humankind on the Devastation of the Environment by Animal Industries and Manufacturing Facilities
- Restore Life to all Waterways and Reforest all Available Land as Public and Global Policy
- Seven Principles for Sustainable, Healthy, and Clean Water Resources
    1. Water is a basic human right.
    2. Water belongs in the public domain.
    3. Water is not a commodity.
    4. Water resources must be protected at all costs.
    5. Dam constructions must be halted as all large dams are deconstructed.
    6. Establish local Water Governance Councils.
    7. Establish a Global Water Convention.
- Harvest and Use Algae, Kenaf, and Hemp Instead of Trees
- Restore Air Quality with Algae Ponds
- Increase and Modernize Public Transportation Systems

Relationships

- Eliminate Capitol Punishment
- Study all Violent Offenders to Determine the Underlying Cause of their Malevolent Behavior
- Decriminalize all Drugs and Set Up Drug Counseling and Education Centers
- Teach Environmental Restoration Skills and Sustainable Farming Skills to all Prison Inmates
- Shelters for the Homeless, Abused, and Elderly
- Free Confidential Birth Control and Education for all People
- Free Education for all to Nourish Artistic Expression, Natural Talents, and Cultural Diversity
- Create Regional Cultural Events
- Protect the Natural Evolution of Indigenous Cultures
- Create Incentives for Communal Groups to Become Self-Sufficient

Commerce and Government

- Cease all Corporate Donations to Elections
- Disallow any Corporate Influence in the Law Making Process
- End all Corporate Subsidies
- Restrict Corporations from Broadcasting Rights and Make Television Responsible to Humanity
- Reduce Global Trade and Eliminate "Free Trade" Agreements
- De-Globalize all Corporate Structures and Regulate Them Locally
- Eliminate the Corporate Assumption of the Status as Constitutionally Protected Human Beings
- Create a Global Reference Currency

- Localize Money Flow by Using a Community Currency
- Halt all Sales and Gifts and Cease Production of all Arms, Munitions, Weaponry, and Devices of Destruction
- Change Election Procedures to Instant Runoff Voting and Offer Incentives to Vote
- Tax only Consumption, Pollution, and Resource Depletion, not Labor or Corporations
- Replace Product Sales Economy with Customer Service Economy
- Invest in Natural Capital
- Create Incentives for all Sustainable Technologies
- Create Peace, Cultural, Health, and Environmental Branches of the World's Governments

Afterword

What Was

We've seen how it was that one particularly unusual culture of humankind came to set itself above the rest of creation, and in so doing, fabricated a wall that grew thicker with time. And with time, we've become separated not only from the rest of creation, but from our own past. Our idea of history covers barely a small glimmer of the human experience, and yet we are allowing the edicts and events that emanate from this sliver of time to determine the very fate of all life.

As our culture tried to separate humanity from the rest of the natural world, we've seen how we became ashamed of anything within us that harked of our true connection to natural processes. We denigrate animals because we need to feel superior to them. We denigrate women because of their procreative functions as they remind mankind of our kinship to the animal kingdom. This denial causes us to be embarrassed of our sexuality and our sexual desires. We denigrate any sex outside of marital sex for reproduction because we need to deny any desire to enjoy our animal bodies. We denigrate the Earth from whence we have come because we believe that humanity belongs to a different reality. Our spirituality is rooted in our imaginations rather than the Earth of which we are made.

We've looked at how ancient hunting/herder peoples birthed the idea of an alien male deity to help them conquer their neighbors, which has perpetuated the intolerance and the aggressive behavior of our culture. And we've examined the Agricultural Evolution that sparked the population explosion and totalitarian society.

What Is

We've seen how our species is annihilating life, and how we are living in historic times—the greatest mass extinction of life that humankind has ever witnessed. We've seen how separated we are from true reality as we retreat into alternate worlds of our own imaginations that serve to further alienate us from nature. And as our culture and all its members make this retreat, our minds become more and more narrowly focused, preventing us from seeing the Truth that could not be more obvious. Each individual knowing so much about so little, yet unable to see the whole reality of what it is we have actually created.

We've seen how our health, Earth's ecosystems, our culture, and our institutions are eroding and will soon crumble. We've examined the old vision of totalitarianism, which bankrupts the quality of life by its propensity to mitigate and control one segment of the population for the exclusive benefit of another. We've also learned that this totalitarian inclination has also had an enormous effect on other life forms. We've further learned that our agrarian cultural background initiated and maintained an unsustainable rate of growth in population at the expense of all of the life systems here on Earth. This is because agrarian values favor large families, a sedentary lifestyle, and continual expansion of food production to support the rapidly expanding population base.

What Will Be

The Cultural Evolution outlined in this book is virtually the complete antithesis of the current paradigm. It's about returning to the true human values living in harmony with the natural world, living humanely, cooperating with each other, and finding our true purpose in life. It's about educating ourselves to be appreciative of the natural cycles and knowing where we fit in the grand scheme. It's about merging with the Universe, valuing the natural world, and tearing down the walls that separate our culture, and ultimately our species, from the beauty of creation.

This cultural vision is about abandoning our quest to expand and exploit, learning to cooperate rather than compete with each other, and restoring that which we have destroyed.

There are a few principles and basic laws set forth in *Cultural Vision* that are worth reviewing:

- **The Principle of the Human Symbiotic Relationship with the Planetary Biosphere**

  Humanity is a symbiotic part of the biosphere that encompasses the planet Earth. The survival of our species is wholly dependant upon the Earth's biosphere. Neither the planet, nor any piece of it, is the exclusive property of Homo sapiens.

- **The Rule of Speciel Equality**

  All species of animals have the entitlement to coexist with all other species. No one species has the right to the wholesale command and rule over another.

- **The Bottom Line**

  If the global Domination culture fails or collapses, our entire species is vulnerable.

- **The Four Kinds of Wealth**

  1. Physical
  2. Natural
  3. Social
  4. Financial

- **The Four Categories of Experience**

  1. Health
  2. Environment
  3. Relationships
  4. Commerce and Government

- **The Three Primary Human Values**

  1. Spiritual Connectedness to the World
  2. Spiritual Connectedness to Each Other
  3. Spiritual Connectedness to Community

- **The Foundational Rule**

  Overpopulation decreases the quality of life. (Humans evolved with band/tribal populations of approximately 30/500.)

- **The Principle of Cultural Non-interference**

  No culture has the right to impose its own morals or beliefs on another or interfere with another culture's evolution. Cultural boundaries are sacred.

I would not want to have any influence on what future tribal cultures might eventually spin out of a cultural evolution such as this; that would be up to those people (provided they adhered to the Principle of Cultural Non-interference). By always proceeding towards diversity and cooperation, we can provide future generations with the tools they need to prosper. With the inevitable collapse of the Domination culture, however, we need to begin the work now of replacing it with something as big (temporarily) based upon human values. This is how we can create a reality big enough to give future generations the ability to choose what is best for them on a healthy planet.

Bibliography

Karen Armstrong, ***A History of God***, New York: Ballentine, 1993
Written by a former Roman Catholic nun, this book examines in great detail the evolution of the human concept of "God," and the power the three great monotheistic religions (Islam, Christianity, and Judaism) wield in our modern reality. Karen Armstrong clearly illustrates how the new concept of a singular male deity emerged from the rampant human confusion that was so prevalent 4,000 years ago. The three great religions were fashioned by their followers to suit their own particular socio-economic needs, particularly those of the ancient rulers.

Maude Barlow and Tony Clarke, ***Blue Gold***, New York: The New Press, 2002
The global water resources are rapidly being depleted and polluted. Transnational corporations are privatizing water supplies and distribution. Governments are redefining the true value of water as they increasingly categorize it as a human *need* and not the basic human *right* it should be. It is quite predictable if we remain on the present course that the wars of the future may be fought over water rights.

Lester R. Brown, ***Tough Choices: Facing the Challenge of Food Scarcity***, Worldwatch Institute, 1996
Our population is growing at a faster rate than the world's agricultural output. These two trends spell trouble. This well-documented book shows that the impact of these diverging trends will be much deeper than we had previously predicted and is a valuable source of indisputable statistics.

William H. Calvin,
A Brain for all Seasons: Human Evolution and Abrupt Climate Change
Chicago: The University of Chicago Press, 2002
Abrupt climate change in which a global ice age could be precipitated in a matter of a few years, as happens every few thousand years. Our ancestors lived through hundreds of such abrupt episodes. Unfortunately, the activities of human industries are causing a gradual warming of the atmosphere that has the potential to trigger a global ice age by shutting down the gulf stream flow in the Atlantic.

Joseph Campbell, *Transformations of Myth Through Time,*
New York: Harper and Row, 1997
This book is a collection of thirteen lectures given by Mr. Campbell. He examines the development of mythologies around the world and the impact these myths have had on the human cultures that invented them. Every culture has devised myths to help explain the mysteries of life. In this book we find out why this is important and what impact this human tendency is having on our own modern culture.

Theo Colborn, Dianne Dumonoski, and John Peterson Myers,
Our Stolen Future, New York: Plume/Penguin, 1997
Man-made chemicals are flooding our environment and disrupting natural processes. Wildlife, human immune systems, and hormonal function are all being denigrated and altered by synthetic compounds never before seen on our planet. The changes that have already occurred and the ones we have yet to discover will profoundly change the reality future generations will have to contend with.

Christian de Duve, *Vital Dust,* New York: BasicBooks, 1995
An excellent scientific explanation of the emergence of life on Earth up to Homo sapiens can be found in this well detailed book. Nobel laureate Professor de Duve offers the most complete yet comprehensible description of the forces of natural evolution I've ever seen.

Nicols Fox, *Spoiled*, New York: Penguin Books, 1998
Changes in how we process our food have allowed pathogens to enter our food supply. Diseased animals and an unbalanced ecology have created a niche for various opportunistic microorganisms to run rampant. These are dangerous times, but mealtimes can be deadly. This comprehensive investigation uncovers the shoddy practices of the organizations to which we have handed over the responsibility of providing our sustenance. It's time to take that responsibility away from the private sector.

Julian Jaynes,
The Origin of Consciousness in the Breakdown of the Bicameral Mind
Boston: Houghton Mifflin Company, 1990
By examining ancient artifacts and writings left behind by humans over the past forty thousand years, we have many clues about how Homo sapiens lived and what was important to them. Most surprising, perhaps, is the discovery that thought pro-

cesses have changed. How we think and what we think about is dictated by the sophistication of the language we have at our disposal, or lack thereof, that we learn early in life. Language, and the level of consciousness that it engenders, evolved long after the first fully modern human walked the Earth. The left and right hemispheres of the human mind have been culturally transformed, and as such have been consciously disconnected, except in the dream state and the mental state of schizophrenia. The bicameral mind, as well as the minds of schizophrenics, receives information coming from the spontaneously information generating right brain which feeds to the left brain whatever data it contains, which it may perceive as being pertinent to the current situation. This information comes from the vast warehouse of past experience that is stored as sounds and images.

When the ancient human was faced with all the things of daily life that required a decision, the answer would come from the right to the left hemisphere as a hallucination, as if the sounds and images were actually being experienced rather than simply remembered. Ancients may have imagined that they were hearing gods, as to them these voices from the past were coming to them from the ether. Now we find that the ancient bicameral mind has broken down in modern civilization, and our culture is run by left-brained thinking. Consciousness as we imagine it to be is not a given, but rather a learned process, as well as an evolutionary one that continues today.

Dr. Bernard Jensen & Mark Anderson, ***Empty Harvest,***
Garden City Park, NY: Avery Publishing, 1990
The link between our industrialized agricultural system, the depleted soils upon which our food is grown, and the deterioration of human health can no longer be debated. Only the world-famous Dr. Jensen can explain these complicated interconnecting life systems in such a comprehendible manner. The bottom line of *Empty Harvest* is that the quality of the food we eat is directly related to the quality of our health.

Bernard Lietaer,
The Future of Money: Creating New Wealth, Work, and a Wiser World
London: Century, The Random House Group Limited, 2001
Bernard Lietaer has been a central banker, a general manager of currency funds, a senior consultant to multinational corporations and developing countries, was picked by *Business Week* as the world's top currency trader, was a consultant for over 12 years to multinational corporations on four continents, as well as a Professor of

International Finance. More information is on his website at www.transaction.net/money.

This book is excellent, and it shows how a sustainable future is attainable—but we need to examine the deepest part of how we relate to each other, both globally as well as within our communities. Money is not the root of all evil, but the kind of money and the manner in which it is traded is the root of all cultures. There are two arenas of trade which have no corresponding currency: global and community. The currencies of our nation-states are not suitable for the exclusive money for international or local trade or transactions.

Based on the four mega-trends of monetary instability, higher percentages of older citizens, the information revolution, and climate change, this book is an examination of four possible scenarios of how current trends could alter the course of history.

1. In "The Corporate Millennium" governments are disbanded, central banks become irrelevant, and the world is run "Big Brother" style by global mega-corporations with their own currencies.

2. In the "Caring Communities" scenario a monetary crash pushes people into small self-sustaining communities that isolate themselves behind gates and walls.

3. "Hell on Earth" illustrates the breakdown that would occur if the current system should break down resulting in an obscene gulf between the rich and the poor.

4. Finally the scenario which he advocates and envisions is called "Sustainable Abundance," where the environment is cared for, the poor and the unemployed are engaged in their communities, the elderly are provided with a high degree of personal care, and the working class has time and fulfillment in their personal lives.

Howard F. Lyman with Glen Merzer, **Mad Cowboy**, New York: Scribner, 1998
Howard Lyman has been on a speaking tour that never ends. He gained national notoriety when he appeared on the Oprah Winfrey television show and offended cattlemen in Texas with the truth about the dangers of eating meat. His book is short and fun to read, while dispensing irrefutable facts on what effects our addiction to flesh is having on the environment and our health.

Burton L. Mack, ***The Lost Gospel, The Book of Q & Christian Origins,***
New York: HarperCollins, 1993
Was Jesus just a man, or was he the "Son of God?" The answer to this question becomes self-evident as Burton Mack examines the first four books of the New Testament. By closely analyzing the time in which each was written and who wrote each one, as well as the stories and the quotations of Jesus, it can be concluded that each author used another book that contained the sayings of Jesus, known today in theological circles as "Q." It becomes clear how this concept of Jesus' fulfilling prophesy as the Christ is a fabrication of the authors of the four gospels, each adding his own embellishments upon the other.

Jerry Mander and Edward Goldsmith, editors,
The Case Against the Global Economy and For *a Turn toward the Local,*
San Francisco: Sierra Club Books, 1996
This book combines the writings of 43 leading economic, agricultural, cultural, and environmental experts who charge that free trade and economic globalization are producing disastrous results on humanity and all life on Earth. Real power has been moved away from citizen democracies and nation states to global corporate bureaucracies. This book makes it abundantly clear that we must reverse our course by turning away from globalization and move instead towards a globally revitalized democracy, local self-sufficiency, and ecological health.

Jerry Mander, ***Four Arguments for the Elimination of Television,***
New York: Quill, 1978
Television has (and still is) altered human culture and society and even our minds. By dictating what we think about, television has become the most powerful tool for shaping public awareness and opinion that has ever been. The mental effects of viewing television are far more than meets the eye. It alters the structure of our minds.

Jerry Mander, ***In Absence of the Sacred***, San Francisco: Sierra Club Books, 1992
Corporations rule. Technology is inherently dangerous. While these statements sound radical, they reflect the only possible conclusion after reading *In Absence of the Sacred*. We have entirely recreated the human experience in the space of just a few years. Isolated and no longer living with a close bond to the Earth, we are drifting further into our imaginations that have no root in Earthly realities. Without this root, our sense of the sacredness of life itself is replaced with a ravenous desire for consumer products that we don't truly need. Through advertising cam-

paigns that can control what the public thinks about, corporations force their agenda into our conscious minds. Jerry Mander shows why we must study what few indigenous cultures are surviving to understand what it is to be a natural human, instead of a programmed one.

Jim Mason,
An Unnatural Order: Why We are Destroying the Planet and Each Other,
New York: Continuum, 1997

How is it that animals and women are considered to be subservient to men in our society, culture, and our own minds? Jim Mason demonstrates the relationship between the reduction of women (misogyny) and the reduction of animals and nature. "Misothery" is the word he coined to mean "hatred of animals." He shows how ancient hunting cults dominated by men laid the foundation that is the base upon which our present day culture is built. He further illustrates how the advent of agriculture created the agrarian cultural value-system of intolerance that we have today.

Daniel Quinn, *Ishmael,* New York: Bantam Books, 1993

Ishmael was the winner of the Turner Tomorrow Fellowship award. In this fictional story of a gorilla, Daniel Quinn shows why we, as a worldwide culture, are killing ourselves along with our mother Earth. He shows that there are several extremely dangerous assumptions we make, based upon our culture, about how life should be lived. The primary faulty premise we make is that earth was created for Man. This sort of belief is not sustainable and must, at some point in the future, break down. In fact, we learn that it already has.

Daniel Quinn, *The Story of "B",* New York: Bantam Books, 1996

In this fictional story of a mysterious underground lecturer and philosopher, Quinn illuminates enormous details of the evolution of human culture, unsustainable agricultural practices, and world religions. I gained many insights to the causes of the violence, hatred, famine, and intolerance found in our history books and still find today. In this book we are introduced to the concept of "totalitarian agriculture," which emerged about ten thousand years ago in what we call the Agricultural Revolution. Daniel Quinn teaches that this is where the root of our problems emanates. We control the planet in an unnatural way, and deny food and habitat to all other life forms, as well as members of our own species.

Paul H. Ray, Ph.D. and Sherry Ruth Anderson, Ph.D., ***The Cultural Creatives,***
New York: Three Rivers Press, 2000
This book is the most optimistic and forward-looking compendiums on the cultural forces which are shaping the future of humankind I've yet to read. The clarity with which they detail the various aspects of the trends which are about to change the way we all live was to me, breathtaking and inspirational. Their conclusions are based upon thirteen years of research involving more than one hundred thousand Americans. This book is a "must read" for any progressive thinking person or group.

Jeremy Rifkin, ***Beyond Beef: The Rise and Fall of the Cattle Culture***,
New York: Plume/Penguin, 1993
Beef has been a driving force in our culture for several millennia, and could be considered a culture unto itself. But few know, or care to know, the devastation of our environment and the threat to human health caused by these creatures. Religions have revolved around them, as have economies. But the degradation of the soils and the pollution of the waterways they cause will soon overwhelm the bio-systems that all life depends upon. Jeremy Rifkin gives us the sociological background behind this industry and makes the solutions seem all too obvious.

John Robbins, ***The Food Revolution,*** York Beach, ME: Conari Press, 2001
In the late 1980s, John Robbins forged the way for vegetarianism to enter the mainstream with his groundbreaking exposé on meat consumption, *Diet for a New America*. In *The Food Revolution,* Robbins covers all the various aspects of our meat-centered culture: the effect on human health and Earth's environment, as well as the moral implications of torturing and murdering defenseless animals for profit. He also demonstrates the reckless disregard today's mega agri-businesses have towards the future of life, and the plight of the small farmer through bio-technology that permeates the industry. He describes the incredulously irresponsible tactics used by corporations to force genetically modified organisms into the food chain as well as our delicate biosphere.

Brian Swimme & Thomas Berry, ***The Universe Story,***
New York: HarperSanFrancisco, 1992
"From the Primordial Flaring Forth to the Ecozoic Era," this is one "story" everyone can enjoy. While *The Universe Story* does contain scientific information, it is presented in terms that make it easy to comprehend. *The Universe Story* traces time from the "Big Bang" to the creation of life on Earth to the modern day

human and the various societies and cultures that have existed up until the present time. It not only makes the truth of our roots as a species evident with absolute clarity; it also demonstrates the human potential to manifest our future.

John Tuxill, ***Losing Strands in the Web of Life: Vertebrate Declines and the Conservation of Biological Diversity***,
Worldwatch Institute, Worldwatch Paper 141, May 1998
The title fairly speaks for itself. There were some tough facts to swallow when I read it, but let's hope that tough facts are followed by decisive actions. We must reverse this trend or we won't be far behind in the mass extinction of life that is occurring on a global basis.

References

Susan Blackmore, *The Meme Machine*, Oxford, NY: Oxford University Press, 1999

Goran Burenhult, General Ed., *The First Humans: Human Origins and History to 10,000 BC*, San Francisco: HarperCollins, 1993

Marija Gimbutas, *The Living Goddesses*, Berkeley, CA: University of California Press, 1999

Marvin Harris, *Cannibals and Kings: The Origins of Cultures*, New York: Random House, 1979

_____, *Cows, Pigs, Wars, and Witches: The Riddles of Culture*, News York: Random House, 1974

_____, *Our Kind*, New York: Harper & Row, 1989

Thom Hartmann, *The Last Hours of Ancient Sunlight*, New York: Three Rivers Press, 1999

_____, *Unequal Protection: The Rise of Corporate Dominance and the Theft of Human Rights*, Rodale, 2002

Paul Hawken, Amory Lovins, and L. Hunter Lovins, *Natural Capitalism*, Boston: Back Bay Books, 1999

Jacquetta Hawkes, *The First Great Civilizations*, New York: Alfred A. Knopf, 1973

The Holy Bible, Authorized (King James) version Derrick Jensen, *A Language Older Than Words*, New York: Context Books, 2000

Richard Leakey and Roger Lewis, *The Sixth Extinction*, New York: Doubleday, 1995

George B. Leonard, *The Transformation: A Guide to the Inevitable Changes in Humankind,* New York: Delacorte Press, 1972

Gary Paul Nabhan, *Cultures of Habitat: On Nature, Culture, and Story,* Washington D.C.: Counterpoint, 1997

John E. Pfieffer, *The Emergence of Man,* New York: Harper & Row, 1972

John Robbins, *Diet for a New America,* Walpole, NH: Stillpoint Publishing, 1987

Ocean Robbins & Sol Solomon, *Choices For Our Future,* Summertown, TN: Book Publishing Company, 1994

Paul Shepard, *The Others: How Animals Made Us Human,* Washington D.C.: Island Press, 1996

Pamela Tuchscherer, *TV Interactive Toys: The New High Tech Threat to Children,* Bend, OR: Pinnaroo Publishing, 1988

Alan Watts, *Nature, Man and Woman,* New York: Pantheon Books, 1969

Endnotes

1. George B. Leonard, 1972, page 41.
2. Marvin Harris, 1979, page 69.
3. Ibid., page 13.
4. Daniel Quinn, 1996, page 253.
5. John E. Pfieffer, 1972, page 275.
6. Ibid.
7. Marija Gimbutas, 1999, pages 112-115.
8. Julian Jaynes, 1990, pages 140-143.
9. Daniel Quinn, 996, page 262.
10. Marvin Harris, 1979, page 29.
11. Daniel Quinn, 1996, page 264.
12. Marvin Harris, 1979, page 4.
13. Ibid.
14. Jeremy Rifkin, 1993, pages 24-26.
15. Julian Jaynes, 1990, pages 293-294.
16. Ibid, pages 238-240.
17. Daniel Quinn, 1996, Page 266.
18. Karen Armstrong, 1993, page 12.
19. Daniel Quinn, 1993, pages 169-170.
20. Julian Jaynes, 1990, page 299.
21. Marvin Harris, 1979, page 57.
22. Daniel Quinn, 1996, Page 268.
23. Burton L. Mack, 1993, pages 3-5.

24. Ibid, page 2.
25. Joseph Campbell, 1997, page 196.
26. Jeremy Rifkin, 1993, pages 22-23.
27. Alan Watts, 1969, page 52.
28. Daniel Quinn, 1996, Page 269.
29. Paul H. Ray and Sherry Ruth Anderson, 2000, pages 27-30.
30. Luther's Works, Volume 47: The Christian in Society IV, Philadelphia: Fortress Press, 1971, p. 268-293.
31. Derrick Jensen, 2000, page 94.
32. Daniel Quinn, 1996, Page 271.
33. Marvin Harris, 1979, page 175.
34. Bernard Lietaer, 2001, pages 50-55.
35. Ibid, page 78.
36. Joel Bleifuss, "Know Thine Enemy," *In These Times,* February 8, 1998, Page 16.
37. Paul H. Ray and Sherry Ruth Anderson, 2000, p. 80-84.
38. Ibid, p. 80-89.
39. Jim Mason, 1997, page 272.
40. IMS Health Report, "Pharmacast and Beyond: A Study of the G7 Antidepressant Market" (1998).
41. Marvin Harris, 1989, page 501.
42. Daniel Quinn, 1996, Page 285-286.
43. Aaron Sachs, "Dying For Oil," *World Watch,* May/June 1996, Vol 9, No 3, page 16.
44. Richard A. Fineberg, "Pipeline in Peril," *Earth Island Journal,* Summer 1998, page 35.
45. "No More Oil!" *Earth Island Journal,* Fall 1998, page 24.
46. Senator Dale Bumpers, "Corporations Get the Gold, Taxpayers Get the Shaft," *Earth Island Journal,* Spring 1998, page 27.

47. "Hands off Public Lands," Zero Cut Support Project, POB 1042, Fall Creek, OR 97438.
48. "Zero Cut on Public Lands," *Earth Island Journal*, Summer 1997, page 8.
49. Curtis Runyan, "Indonesia's Discontent," *World Watch*, May/June 1998, Vol.11, No. 3, page 17.
50. Robert Bryce, "Spinning Gold," *Mother Jones*, September/October 1996, page 66.
51. Janet Abramovitz, "Nature's Hidden Economy," *World Watch*, January/February 1998, Vol. 11, No. 1, page 13.
52. Howard F. Lyman with Glen Merzer, 1998, page 123.
53. Peter Goin, 1996.
54. George Wuerthner, "Why Healthy Forests Need Dead Trees," *Earth Island Journal*, Fall 1995, page 22.
55. Susque Hannah, "Clearcut Landslides Devastate West Coast," *Earth First! Journal*, February-March 1997, p. 5.
56. Paul Hawken, "Natural Capitalism," *Mother Jones*, March/April 1997, page 43.
57. Chad Hanson, "The National Forest Rip-Off," *Earth Island Journal*, Fall 1998, page 11.
58. "Hands off Public Lands," Zero Cut Support Project, POB 1042, Fall Creek, OR 97438.
59. Ocean Robbins & Sol Solomon, 1994, page 36.
60. Ibid, page 37.
61. Howard F. Lyman with Glen Merzer, 1998, page 42.
62. Ocean Robbins & Sol Solomon, 1994, page 75.
63. Ibid, page 77.
64. Ibid.
65. Will Nixon, "rainforest shrimp," *Mother Jones*, March/April 1996, page 33.
66. Ibid.

67. Theo Colborn, Dianne Dumonoski, and John Peterson Myers, 1997, page 218.
68. Sara Chamberlain, "Golf Endangers Hawai'ian Ecology and Culture," *Earth Island Journal*, Summer 1995, page 21.
69. Ibid.
70. "If You Can't Beat 'Em, Join 'Em," *Earth Island Journal*, Summer 1995, page 21.
71. Jeremy Rifkin, 1993, pages 186.
72. Howard F. Lyman with Glen Merzer, 1998, page 136-137.
73. Ibid.
74. "Hands off Public Lands," Zero Cut Support Project, POB 1042, Fall Creek, OR 97438.
75. Jeremy Rifkin, 1993, page 216.
76. Ibid, page 213.
77. Ibid, page 215.
78. Ibid, page 217.
79. Ibid, page 216.
80. Ibid, page 210.
81. Ibid, page 221.
82. Ibid.
83. Jim Mason, "Fowling the Waters," *E magazine*, September/October 1995, page 33.
84. Howard F. Lyman with Glen Merzer, 1998, page 13.
85. Jeremy Rifkin, 1993, pages 221.
86. Howard F. Lyman with Glen Merzer, 1998, page 141.
87. Joel Bleifuss, "Radioactive Sludge," *In These Times*, April 28, 1997, page 12.
88. Maude Barlow and Tony Clarke, 2002, page 52.
89. Ibid, page 28.
90. Richard Leakey and Roger Lewis, 1995, page 234.

91. Janet N. Abramovitz, Worldwatch Paper 128, *Imperiled Waters, Impoverished Future: The Decline of Freshwater Ecosystems*, Worldwatch Institute, 1996.
92. Chris Bright, *World Watch*, July/August 1998, Volume 11, No. 4, page 39.
93. Maude Barlow and Tony Clarke, 2002, page 48—50.
94. Lester R. Brown, Gary Gardner, and Brian Halweil, *Beyond Malthus: Sixteen Dimensions of the Population Problem*, Worldwatch Paper 143, September 1998, page 16.
95. "A Mass for Rain, A Cry to God," and "Water: In Short Supply," *Earth Island Journal*, Fall 1998, page 22.
96. Lester R. Brown, Gary Gardner, and Brian Halweil, *Beyond Malthus: Sixteen Dimensions of the Population Problem*, Worldwatch Paper 143, September 1998, page 39.
97. Anne Platt McGinn, "Freefall in Global Fish Stocks," *World Watch*, May/June 1998, Vol 11, No. 3, page 10.
98. Lester Brown, et al., *Vital Signs: 1994*, Worldwatch Institute, 1994, page 32.
99. Carl Safina, "The World's Imperiled fish," *Scientific American*, Nov 1995.
100. Jeffrey St. Clair, "Fishy business," *In These Times,* May 26, 1997, page 14.
101. Seth Borenstein, "Study: Trawling destroying ocean floors," *Knight Ridder Newspapers, Las Vegas Review-Journal,* December 20, 1998, page 17B.
102. Steve Lustgarden, "Fish: What's the Catch?, *EarthSave*, Spring 1996, Vol. 7, Number 1, page 2.
103. Ibid.
104. Will Nixon, "rainforest shrimp," *Mother Jones*, March/April 1996, page 33.
105. Mandi Billinge, "Bay Area Kids Battle Pollution," *Earth Island Journal,* Spring 1998, page 7.
106. Derek M. Brown, "Pfiesteria, How the Meat Industry Destroys Waterways," *Good Medicine*, Winter 1998, Volume VII, Number I, page 16, Physicians Committee for Responsible Medicine.

107. Jonathan Tolman, "Poisonous Runoff from Farm Subsidies," *Wall Street Journal*, September 8, 1995, A10.

108. "Ozone Hole Matches Record," *World Watch*, March/April 1997, Vol.10, No. 2, page 7.

109. "The Ozone Hole Just Got Deeper," *Earth Island Journal*, Winter 1997-98, page 13.

110. Jim Scanlon, "Silenced Science: Arctic Ozone Loss," *Earth Island Journal*, Fall 1998, page 23.

111. Gar Smith, "Oil Spills in the Sky," *Earth Island Journal*, Summer 1997, page 34.

112. Lester R. Brown, Gary Gardner, and Brian Halweil, *Beyond Malthus: Sixteen Dimensions of the Population Problem*, Worldwatch Paper 143, September 1998, page 22.

113. William H. Calvin, 2002, page 315.

114. Ibid.

115. Gary Gardner, When Cities Take Bicycles Seriously, *World Watch*, September/October 1998, Vol.11, No. 3, page 17.

116. Ocean Robbins & Sol Solomon, 1994, page 102.

117. Ibid, page 81.

118. Charlotte Huff, "The Invisible Poison," *Las Vegas Review-Journal*, February 3, 1997, page 1A.

119. *Earth Island Journal*, Summer 1998, page 25.

120. Adam T. Williams, "Paper and Global Warming," *Earth Island Journal*, Summer 1998, page 27.

121. Peter Goin, *Humanature*, Harrisonburg, VI: University of Texas Press, 1996, page 1.

122. William H. Calvin, 2002, page 315.

123. Christopher Flavin, "Last Tango in Buenos Aires," *World Watch*, November/December 1998, Vol 11, No. 6, page 8.

124. "El Nino? Meet El Papa!", *Earth Island Journal*, Fall 1998, page 13.

125. "Ed Ayres, "Global Temperature Reaches Historic Highs," *World Watch*, November/December 1998, Vol 11, No. 6, page 8.
126. "As the World Burns," *Earth Island Journal*, Fall 1998, page 23.
127. "The South Pole is Melting," *Earth Island Journal*, Spring 1998, page 3.
128. Molly O'Meara, "Antarctica Ice Shelf Crumbling," *World Watch*, July/August 1998, Vol 11, No. 4, page 8.
129. Andy Caffrey, "Antarctica's 'Deep Impact' Threat," *Earth Island Journal*, Summer 1998, page 26.
130. William H. Calvin, 2002, page 315.
131. Molly O'Meara, "The Risks of Disrupting Climate," *World Watch*, November/December 1997, Vol. 10, No. 6, page 18.
132. William H. Calvin, 2002, page 315.
133. Intergovernmental Panel on Climate Change (IPCC), *Climate Change 1995: Impacts, Adaptations and Mitigation of Climate Change.*
134. Molly O'Meara, "The Risks of Disrupting Climate," *World Watch*, November/December 1997, Vol. 10, No. 6, page 13.
135. Ibid.
136. William H. Calvin, 2002, page 315.
137. Ibid, page 19.
138. IPCC, *Climate Change 1995: Impacts, Adaptations and Mitigation of Climate Change.*
139. "Global Warming/Greening," *Earth Island Journal*, Summer 1997, page 14.
140. William H. Calvin, 2002, page 315.
141. "Life in a Global Greenhouse," *Earth Island Journal*, Winter 1997-98, page 33.
142. Christopher Flavin, "Not Whether, but *Weather*," *World Watch*, May June 1996, Vol 9, No. 3, page 2.
143. "Study: Human to blame in record year for weather damage," Associated Press, *Las Vegas Review-Journal*, November 28, 1998, 11A.
144. Ibid.

145. William H. Calvin, 2002.
146. Richard Leakey and Roger Lewis, 1995, page 241.
147. Lester R. Brown, Gary Gardner, and Brian Halweil, *Beyond Malthus: Sixteen Dimensions of the Population Problem,* Worldwatch Paper 143, September 1998, page 19.
148. Richard Leakey and Roger Lewis, 1995, pages 242-243.
149. "Study: Earth's resources declining," Associated Press, October 1, 1998.
150. John Tuxill, *Losing Strands in the Web of Life: Vertebrate Declines and the Conservation of Biological Diversity,* Worldwatch Institute, Worldwatch Paper 141, May 1998, page 17.
151. Ibid, page 21.
152. Theo Colborn, Dianne Dumonoski, and John Peterson Myers, 1997, page 160.
153. Gary Paul Nabhan, 1997, page 34.
154. "Sea life flees warmer Pacific—or dies," *USA Today,* Associated Press, July 10, 1998.
155. Molly O'Meara, "The Risks of Disrupting Climate," *World Watch,* November/December 1997, Vol. 10, No. 6, page 18.
156. John Tuxill, *Losing Strands in the Web of Life: Vertebrate Declines and the Conservation of Biological Diversity,* Worldwatch Institute, Worldwatch Paper 141, May 1998, page 33.
157. Gray Brechin, "How Paradise Lost," *Mother Jones,* November/December 1996, page 42.
158. Paul and Anee Ehrlich, *Healing the Planet,* Addison Wesley, 1991.
159. Howard F. Lyman with Glen Merzer, 1998, page 142.
160. Ashley Mattoon, "Plants in Peril," *World Watch,* July/August 1998, Vol.11, No.4, page 9.
161. "World's Plants in Danger," *Earth Island Journal,* Summer 1998, page 16.
162. "Extinctions R Us," *Earth Island Institute,* Summer 1996, page 14.
163. Ed Ayres, "The Fastest Mass Extinction in Earth's History," *World Watch,* September/October, 1998, Vol 11, No. 5, page 6.

164. Steve Lustgarden, "Fish: What's the Catch?, *EarthSave*, Spring 1996, Vol. 7, Number 1, page 1.
165. Jeremy Rifkin, 1993, pages 226-229.
166. "Last Decade for World's Forests?", *Earth Island Journal*, Summer 1998, page 3.
167. Jim Scanlon, "Silenced Science: Arctic Ozone Loss," *Earth Island Journal*, Fall 1998, page 23.
168. Gary Paul Nabhan, 1997, page 269.
169. Cindy Durehring, "A Beacon in a Toxic Storm," *Earth Island Journal*, Spring 1998, page 34.
170. Theo Colborn, Dianne Dumonoski, and John Peterson Myers, 1997, page 90.
171. John Robbins, 1987, pages 329-331.

    Theo Colborn, Dianne Dumonoski, and John Peterson Myers, 1997, pages 26-28.
172. Dr. Bernard Jensen & Mark Anderson, 1990, p 129.
173. Patrick Wright, Ph.D., *Grains & Greens*, Wright Publications (1990), page 38.
174. Dr. Bernard Jensen & Mark Anderson, 1990, p 129.
175. Richard G. Foulkes, MD and Anne C. Anderson RPN, "Impact of Artificial Fluoridation on Salmon in the Northwest USZ and British Columbia," Special 16-page Report—Fluorides and the Environment, page 8-9, *Earth Island Journal*, Summer 1998, center section.
176. Dr. Bernard Jensen & Mark Anderson, 1990, p 131.
177. Leon Chaitow and Natasha Trenev, *Probiotics*, Thorsons (1990), page 90.
178. Theo Colborn, Dianne Dumonoski, and John Peterson Myers, 1997, pages 112-121.
179. Anne Schonfield, MBAN, "Campaign Against Methyl Bromide, Ozone-Killing Pesticide Opposed," *Earth Island Journal*, Summer 1995, page 19.
180. Jennifer D. Mitchell, "Nowhere to Hide, The Global Spread of High-Risk Synthetic Chemicals," *World Watch*, March/April 1997, Vol. 10, No. 2, page 29.

181. Ibid, pages 29-31.
182. Theo Colborn, Dianne Dumonoski, and John Peterson Myers, 1997, page 113.
183. Ibid, pages 47-67, 72.
184. Ibid, page 68-75.
185. Jennifer D. Mitchell, "Nowhere to Hide, The Global Spread of High-Risk Synthetic Chemicals," *World Watch,* March/April 1997, Vol. 10, No. 2, page 31.
186. Theo Colborn, Dianne Dumonoski, and John Peterson Myers, 1997, pages 172-173.
187. Jennifer Myers, "Nations Plan Phase-Out of Deadly Chemicals," *World Watch,* September/October 1998, Vol 11, No. 5, page 7.
188. Theo Colborn, Dianne Dumonoski, and John Peterson Myers, 1997, page 167.
189. Gary Paul Nabhan, 1997, page 213.
190. Ocean Robbins & Sol Solomon, 1994, page 74.
191. John Robbins, 1987, page 357.
192. Ibid, page 358.
193. Dr. Bernard Jensen & Mark Anderson, 1990, page 32.
194. Ibid, page 59.
195. Ibid, page 55-58.
196. Ocean Robbins & Sol Solomon, 1994, page 89.
197. Howard F. Lyman with Glen Merzer, 1998, page 21.
198. Nicols Fox, *Spoiled*, 1998, page 151.
199. Ibid, page 168.
200. Ibid, page 193.
201. Howard F. Lyman with Glen Merzer, 1998, page 38.
202. Nicols Fox, 1998, page 193.
203. Ibid, page 179 and John Robbins, 1987, pages 302-303.
204. Nicols Fox, 1998, page 86.

205. Ibid, page 252.
206. Ibid.
207. Ibid, page 194.
208. Howard F. Lyman with Glen Merzer, 1998, pages 38-39.
209. Nicols Fox, 1998, pages 245-246.
210. John Robbins, 1987, pages 303-4.
211. Nicols Fox, 1998, page 56.
212. Ibid, page 158.
213. Ibid, page 199.
214. Howard F. Lyman with Glen Merzer, 1998, page 40.
215. Ibid.
216. "Is Our Fish Fit to Eat?", *Consumer Reports*, February 1992.
217. *Good Medicine*, Summer 1996, Volume V, Number 3, page 12, Physicians Committee for Responsible Medicine.
218. Nicols Fox, *Spoiled*, 1998, page 250.
219. Ibid, page 214.
220. Ibid, page 251.
221. Jeremy Rifkin, 1993, page 141.
222. Nicols Fox, 1998, page 251.
223. Jeremy Rifkin, 1993, page 138.
224. Ibid, page 143.
225. Ibid, page 140.
226. Ibid.
227. Nicols Fox, 1998, page 192.
228. Ibid, page 87.
229. John Robbins, 1987, page 111.
230. Ibid, page 112.
231. Ibid, page 111.

232. Nicols Fox, 1998, page 156 and John Robbins, 1987, pages 135-137.

233. "What About Turkey?", *EarthSave,* Fall 1997, Volume 8, Number 3, page 13.

234. Peggy L. Carlson, M.D., "The Failure of Animal Experiments, Part II," *Good Medicine*, Summer 1996, Volume V, Number 3, page 7, Physicians Committee for Responsible Medicine "New Cell Tests Beat Animal Tests," *Good Medicine*, Spring 1997, Volume VI, Number 2, page 7, Physicians Committee for Responsible Medicine.

235. Patrick Wright, Ph.D., *Grains & Greens,* Wright Publications (1990), page 187.

236. Ibid, page 195.

237. Dr. Bernard Jensen & Mark Anderson, 1990, page 37.

238. Howard F. Lyman with Glen Merzer, 1998, pages 25-27.

239. Patrick Wright, Ph.D., *Grains & Greens,* Wright Publications (1990), page 131-132.

240. Ibid, page 180.

241. Ibid, page 177.

242. John Robbins, 1987, pages 258-260.

243. Ibid, page 196.

244. Howard F. Lyman with Glen Merzer, 1998, page 35.

245. Annemarie Colbie, *Food and Healing,* U.S.: Ballentine Books, 1996, page 51.

246. Robert Cohen, "Milk: The Deadly Poison," *Earth Island Journal,* Winter 1997-98, page 19.

247. Ibid.

248. John Robbins, 1987, pages 293-297.

249. Ibid, pages 290-292.

250. Ibid, page 264, and Outwater J, Nicholsons A, Barard ND, Breast cancer and dairy product consumption, Med Hypoth 1997, 6:453-62, and Epidemiology, July 1994, 5(4): 391-7.

251. Risch HA, Jain M, Marrett LD, Howe GR. Dietary fat intake and risk of epithelial ovarian cancer. J National Cancer Inst 1994; 86:1409-15, and Journal of the National Cancer Institute 1994; 86:1409-1415.

252. John Robbins, 1987, pages 270-272.

253. Howard F. Lyman with Glen Merzer, 1998, page 35, and John Robbins, 1987, pages 275-278, and Nicholson AN, et al., The effect of a low-fat, unrefined vegan diet on non-insulin-dependent diabetes mellitus, 1997, in press.

254. John Robbins, 1987, pages 287-290.

255. Ibid, page 300.

256. "Diet and Diverticular Disease," *Good Medicine*, Spring 1995, Volume IV, Number 1, page 5, Physicians Committee for Responsible Medicine.

257. Steve Lustgarden, "Factory Farm Alarm," *EarthSave*, winter 1997, Volume 8, Number 4, page 7.

258. Howard F. Lyman with Glen Merzer, 1998, page 30, and John Robbins, *Diet for a New America*, 1987, page 293.

259. Howard F. Lyman with Glen Merzer, *Mad Cowboy*, 1998, page 34-35, and "Another Strike Against Animal Protein," *Good Medicine*, Autumn 1994, Volume III, Number 4, page 4, Physicians Committee for Responsible Medicine.

260. Meat Linked to Common Blood Cancer, *Good Medicine*, Summer 1996, Volume V, Number 3, page 5, Physicians Committee for Responsible Medicine, and Journal of the American Medical Association, 1996; 275:1315-1321.

261. John Robbins, 1987, page 293

    "Protein and Kidney Cancer," *Good Medicine*, Autumn 1994, Volume III, Number 4, Physicians Committee for Responsible Medicine.

262. Ocean Robbins & Sol Solomon, 1994, page 35.

263. *Good Medicine*, Summer 1997, Volume VI, Number 3, page 8, Physicians Committee for Responsible Medicine.

264. Uncle Sam World's Arms Merchant Again; #53, Arms Trade Oversight Project, Council for a Livable World, August 20, 2001.

265. A Risky Business; U.S. Arms Exports To Countries Where Terror Thrives, Center for Defense Information, November 29, 2001.
266. www.amnesty.org.uk/action/camp/saudi/repression.shtml.
267. Chapter summary from the SIPRI Yearbook 2002: Armaments, Disarmament and International Security (Oxford: Oxford University Press, 2002).
268. Ocean Robbins & Sol Solomon, 1994, page 129.
269. Nicholas Lenssen and Christopher Flavin, "Meltdown," *World Watch*, May/June 1996, page 23.
270. CO, "Mobile Chernobyls," *Earth Island Journal*, Summer 1996, page 13.
271. Jeremy Rifkin, 1993, page 149.
272. Ibid, pages 148-149.
273. Ibid, page 193.
274. Ibid, pages 192-193.
275. Ibid, page 150.
276. Odil Tunali, "Habitat II: Not Just Another 'Doomed Global Conference'," *World Watch*, May/June 1996, Vol. 9, No. 3, page 32.
277. Lester R. Brown and Brian Halweil, "China's Water Shortage Could Shake World Food Security," *World Watch*, July/August, Vol. 11, No. 4, page 10-21. Ocean Robbins & Sol Solomon, 1994, page 82.
278. *World Watch*, May/June 1998, Vol. 11, No. 3, page 39.
279. Judith Bruce, *Families in Focus*, New York: Population Council Publications, 1995.
280. Thom Hartman, 2002, page 193.
281. Thalif Deen, "DRUGS: Global Drug Trade Reaches Staggering Proportions" UNITED NATIONS, Mar 2, 2002 (IPS).
282. Paul Hawken, "Natural Capitalism," *Mother Jones*, March/April 1997, page 46.
283. Salim Muwakkil, "Our National Epidemic," *In These Times*, December 27, 1998, page 3.

284. Paul Hawken, "Natural Capitalism," *Mother Jones,* March/April 1997, page 46.
285. Avery F. Gordon, "Globalism and the Prison Complex: An Interview with Angela Davis," *Race and Class: The Threat of Globalism,* October 1998—March 1999.
286. Thom Hartman, 2002, page 192.
287. Ibid.
288. Russell Mokhiber, "Underworld, U.S.A.," *In These Times,* April 1, 1996, page 14.
289. *Pulling Apart: A State-by-State Analysis of Income Trends,* Center on Budget and Policy Priorities and the Economic Policy Institute (2000).
290. Kevin Quinn, Abt Associates, Inc., Cathy Schoen, The Commonwealth Fund, and Louisa Buatti, Abt Associates, Inc., HEALTH INSURANCE ON THEIR OWN: Young Adults.

    Living Without Health Insurance, May 2000, Most data in this report come from either the March 1999 Current Population Survey (CPS) or The Commonwealth Fund 1999 National Survey of Workers' Health Insurance. The CPS is undertaken monthly by the U.S. Bureau of the Census.
291. Both the *Fortune* and *Wall Street Journal* references as quoted in *Netview,* (Global Business Network News) Vol. 7, No. 1 (Winter 1996) respectively, page 16 and page 9.
292. Bernard Lietaer, 2001, page 132.
293. Aaron Sachs, "Dying For Oil," *World Watch,* May/June 1996, Vol 9, No. 3, page 13.
294. United Nations Research Institute for Social Development, *States of Disarray: The Social Effects of Globalization,* London: UNRISD, 1995.
295. *The Times,* December 17, 2001.
296. Thom Hartman, 2002, page 149.
297. Noreena Hertz, 2001, page 8.
298. John E Pfeiffer, 1972, page 375.
299. George B. Leonard, 1972, page 83.

300. Ibid, pages 35-36.

301. Inspired by Larry Harvey, founder of the Burning Man festival.

302. Ibid, (This entire subsection is based upon several lectures by Mr. Harvey.)

303. George Gerbner, "The Stories We Sell, or Why We Need the People's Communication Charter," *Chicago Media Watch*, Fall 1997, page 3.

304. Bernard Lietaer, 2001, pages 46, 304-305.

305. Waxmann and Hinderliter: *A Status Report on Hunger and Homelessness in American's Cities*: 1996, (US Conference of Mayors, 1520 Eye St. NW, Suite 400, Washington DC 20006-4005).

306. Bernard Lietaer, 2001, page 276.

307. Ibid, page 242.

308. Helena Norberg-Hodge, *The Case Against the Global Economy*, "Shifting Direction," San Francisco: Sierra Club Books, 1996.

309. Thom Hartmann, 2002, page 204.

310. "Feeding Mammon—World Trade Organization: Special Report," *Guardian*, November 30, 1999.

311. Ocean Robbins & Sol Solomon, 1994, page 15.

312. Susan Blackmore, 1999, pages 30-31.

313. Ibid, page 50-51.

314. Ibid, page 40-41.

315. Inspired by Larry Harvey, founder of the Burning Man festival, from a lecture on September 6, 1997.

316. Jim Mason, 1997, page 298.

317. Joseph Campbell, 1972, pages 214—217.

318. George B. Leonard, 1972, page 138.

319. Paul H. Ray and Sherry Ruth Anderson, 2000.

320. Ibid.

321. Co-op America, Harwood Group.

322. David R. Brower, "CPR for Business and the Planet," *Earth Island Journal*, Fall 1998, special center section—"David Brower on the Blue Planet."
323. Gary Paul Nabhan, 1997, page 319.
324. William H. Calvin, 2002, pages 279-280.
325. Brad Lemley, "Anything into Oil," *Discover*, May, 2003, page 50.
326. Marvin Harris, 1989.
327. "Riverbanking Clean Water," *Earth Island Journal*, Fall 1998, page 22.
328. Ibid.
329. George Staples, "Optimal Health Journal Interviews Algae expert George Staples," *The Experts' Optimal Health Journal, Vol 1, Issue 2*, Apprise Publishing, 1997.
330. Howard Gardner, *Frames of Mind: The Theory of Multiple Intelligences*, New York: Basic Books, 1983.
331. Thom Hartmann, 2002.
332. Bernard Lietaer, 2001, pages 249-257.
333. Ibid, page 257.
334. Ibid, page 177.
335. Ibid, page 160.
336. Ibid, page 144-145.
337. Ibid, page 147.

0-595-29146-5

Printed in the United States
17301LVS00004B/207